TEXT RETRIEVAL AND FILTERING:
Analytic Models of Performance

THE KLUWER INTERNATIONAL SERIES ON INFORMATION RETRIEVAL

Series Editor

W. Bruce Croft

University of Massachusetts
Amherst, MA 01003

Also in the Series:

MULTIMEDIA INFORMATION RETRIEVAL: Content-Based Information Retrieval from Large Text and Audio Databases
by Peter Schäuble
ISBN: 0-7923-9899-8

INFORMATION RETRIEVAL SYSTEMS
by Gerald Kowalski
ISBN: 0-7923-9926-9

CROSS-LANGUAGE INFORMATION RETRIEVAL
edited by Gregory Grefenstette
ISBN: 0-7923-8122-X

TEXT RETRIEVAL AND FILTERING:
Analytic Models of Performance

by

Robert M. Losee
University of North Carolina
Chapel Hill, NC   USA

KLUWER ACADEMIC PUBLISHERS
Boston / Dordrecht / London

Distributors for North, Central and South America:
Kluwer Academic Publishers
101 Philip Drive
Assinippi Park
Norwell, Massachusetts 02061 USA

Distributors for all other countries:
Kluwer Academic Publishers
Distribution Centre
Post Office Box 322
3300 AH Dordrecht, THE NETHERLANDS

Library of Congress Cataloging-in-Publication Data

A C.I.P. Catalogue record for this book is available
from the Library of Congress.

---

*The publisher offers discounts on this book when ordered in bulk quantities. For
more information contact: Sales Department, Kluwer Academic Publishers,
101 Philip Drive, Assinippi Park, Norwell, MA 02061*

Printed on acid-free paper.

Printed in the United States of America

Contents

Preface

A scientist commonly professes to base his beliefs on observations, not theories..... I have never come across anyone who carries this profession into practice.... Observation is not sufficient.... theory has an important share in determining belief.

—Arthur Eddington, astronomer

Information retrieval scholars have long studied retrieval performance experimentally. While document ranking procedures consistent with formal models have been developed (Maron & Kuhns, 1960; Robertson & Sparck Jones, 1976; Robertson, 1977; Kraft & Bookstein, 1978; Bookstein, 1983; Cooper, 1995), little work has been done showing how combining a given ranking procedure, query, and database results in a particular level of performance. In this book we describe means by which information retrieval may be studied analytically, allowing one to *describe* current performance, *predict* future retrieval performance and to *understand why* systems perform as they do. Specific means for computing the expected performance are developed, having at their core the *average search length* performance measure, the expected position of a relevant document in a ranked list of documents. This measure has an advantage over many other performance measures in that it is relatively easy to predict.

This work focuses on the performance of systems that retrieve natural language text, considering full sentences as well as phrases and individual words. The last chapter explicitly addresses how grammatical constructs and methods may be studied in the context of retrieval or filtering system performance. The book builds toward solving this problem, although the material in earlier chapters is as useful to those addressing non-linguistic, statistical concerns, as it is to linguists. Those interested in grammatical information should be cautioned to carefully examine earlier chapters, especially Chapters 7 and 8, which discuss purely statistical relationships between terms, before moving on to Chapter 10, which explicitly addresses linguistic issues.

Unambiguous statements of the conditions under which one method or system will be more effective than another are developed. After a simple measure

of performance is developed, techniques for predicting performance are proposed, first for single term queries, and then for multiple term queries, with
and without linguistic knowledge. Many of the most important results are
presented in the text as theorems, with supporting arguments that the author
hopes will be convincing to the reader, while avoiding some details one would
find in more formal mathematical proofs. Corollaries provide assertions that
may be less important than the theorems on which they are based. We also
provide conjectures that the author believes to be true but that may need
both theoretical and experimental results if they are to be generally accepted.
Together, these techniques and results provide tools for studying individual
systems or comparing the performance of different systems.

The best way to learn the analytic techniques that serve as the basis for
these text filtering and retrieval models is through manipulating the models,
teaching one the techniques and assumptions underlying them. This may be
done through the book's exercises and by paper-and-pencil work with the models. Symbolic math packages, such as Mathematica, Maple, and Macsyma,
can be used to develop both symbolic and numeric solutions to more complex
equations, enabling one to expand a model or to verify it using an experimental
database. Multi-term analytic models can become quite large, and coding them
using symbolic mathematics packages on a computer or a high-end calculator
may lead the reader to a deeper understanding of these models than might
be obtained through arduous and error-prone paper-and-pencil manipulations.
These computer packages are useful learning and computing tools and have
been used in the production of a few of the more complex analytic formulae in
the book, as well as all of the graphics.

We provide a number of exercises of varying degrees of difficulty. Some,
marked [*Easy*], are relatively easy and normally will require a few minutes worth
of effort to answer. **All exercises marked [*Easy*] should be completed by
the reader** who wishes to fully integrate the material in the preceding section.
Other exercises, marked [*Moderate*], are of moderate difficulty and might take
ten minutes to an hour or two for most people to solve. More difficult questions are marked with [*Difficult*] and may take significantly more effort and
require more insight than most homework problems. Research problems are
marked [*Research Problem*]. These research exercises are significant problems
that can be answered analytically using methods described here and whose
answers would advance knowledge about text filtering and retrieval. Scholars
reading this work should examine each research problem and consider it, however briefly, before moving on to greener pastures. We have also included some
exercises that examine experimentally those aspects of retrieval and filtering
where knowledge can be gained only through the gathering of empirical data.

To Caitlyn, who doesn't scream
as loudly now when she sees a
Mathematica plot.

Acknowledgments

Several students taking retrieval courses at the University of North Carolina at Chapel Hill provided suggestions about earlier drafts, including Patrick Howell, Heather Chapman, Dawn Sanks, and Sharon Bortz. Graduate assistants Jordan T. Davis, Alenka Sauperl, and Don Sechler provided stylistic comments and corrections on earlier drafts.

Casual conversations with Bill Shaw about retrieval would often include the author being asked whether the analytic model could address a problem or address it in a particular way; Bill's questions inspired some of what is below. I would also like to thank Jerry Saye and Diane Sonnenwald for many discussions about retrieval, education, and the publication of this work, as well as Michael Littman, Greg Newby, and Charlie Viles, who made influential retrieval comments at some point while I was writing. As always, my greatest appreciation goes to my wife and daughter, who endured my doodling, pacing, writing, and editing, without complaint.

1 INTRODUCTION

The future truths of physics are to be looked for in the sixth place of decimals.
—Albert Michelson, U. of Chicago, 1894.

The keywords approach with statistical techniques has reached its theoretical limit and further attempts for improvement are ...... a waste of time
—Sembok and Van Rijsbergen, 1990

1.1 TEXT FILTERING AND RETRIEVAL

The performance of information retrieval and text filtering systems may be understood better and estimated more accurately using formal analytic methods than through the use of experimental tests. Document filtering and retrieval systems compare expressed information needs and any available information sources and then present the best documents to the searcher for possible use. Explaining how such a system works, as well as describing and predicting system performance, is the goal of this book.

These systems provide for the retrieval or identification of information, given a query expressing the user's knowledge about her or his information needs. While these search statements don't always reflect what the user really needs (Wood, Ford, Miller, Sobczyk, & Duffin, 1996), what the query provides can serve as a surrogate for some state of want within the user's brain. These statements of need are expressed in a wide variety of forms: e.g., as statements in a Boolean language about diseases affecting pet fur,

(dogs or cats) and fur and diseases,

or expressions allowing for the weighting of terms,

dogs (.7), cats (.6), fur (.35), diseases (.8),

or statements about information needs expressed in a human language,

I am looking for documents about diseases affecting the fur of house pets such as dogs and cats.

Based on the query presented to the system, documents are retrieved for presentation to the searcher. These documents may be judged as *relevant* or *non-relevant* by the user, or a more varied set of relevance judgements may be provided. User-supplied relevance judgements may be used by the system to increase its knowledge about the characteristics of relevant and non-relevant documents, improving the ability of the system to separate the relevant documents from the non-relevant documents. The performance of filtering and retrieval systems may then be measured by studying the degree to which relevant and more beneficial documents are moved toward the front of an ordered list of documents, to be presented in ranked order to the user.

Using experimental techniques, methods may be learned for improving document ranking, as well as better methods for estimating the characteristics of relevant and non-relevant documents and for indexing the contents of documents. By using large sets of experimental results, one may conclude that a specific method is better than another, over a wide range of environments. The method then might be claimed to be universally superior. Using this *synthetic* approach to studying retrieval, one can build a system likely to be superior to systems built using techniques that have less experimental support.

While experimentation has been the dominant paradigm for studying retrieval and filtering systems, formal *analytic* models have many applications that can lead to an increase in our understanding of these systems (Sebastiani, 1998). When the assumptions of an analytic prediction model are met, and the model accurately describes the characteristics of the retrieval or filtering environment, accurate descriptions and predictions will be obtained. Such analytic methods provide tools useful for determining under what conditions a system consistent with one model of information retrieval or filtering will be superior to a different system. These analytic methods scale well, providing equally accurate predictions for small experimental databases of hundreds of documents and for commercial systems containing hundreds of millions of documents. The latter are particularly difficult, time consuming, and expensive to study with experimental methods. Yet, they are easy to study analytically. Finally, and possibly most importantly, analytic methods allow one to *understand why a particular system performs as it does.*

Experimental studies of retrieval performance provide limited information about specific retrieval methods and phenomena. Simulations using experimental databases only produce new information about the combination of the ranking algorithm, queries, documents, and relevance judgements that occurred together. It can be very difficult to isolate the key factors in the inputs to these simulations to determine what caused the observed results or the exact nature of the relationships between input factors and performance results.

How does one learn, conceptualize, and use methods in practical situations? Consider a situation where one plans on traveling 60 kilometers per hour for 2 hours. How far would one go? Most people will multiply the 60 kilometers per hour by 2 hours, yielding 120 kilometers. This is an analytic approach to solving a "time multiplied by rate equals distance" problem. An experimental approach would be to try driving at 60 kilometers per hour for two hours several times, and then average the distances from each of the drives. One might also drive for 6 seconds several times, and then extrapolate this to the problem of driving for two hours. While this situation may seem overly simple, it appears simplistic only because we *know* how to multiply rate by time to produce distance. Unfortunately, most text filtering and retrieval studies are still at the experimental stage. Analytic models capable of providing the sort of "rate multiplied by time" rules that are needed to understand, predict, and describe the performance of text filtering and retrieval systems are given below, and represent the focus of this book.

An *understanding* of the key factors in the operation of text filtering and retrieval systems serves as the core of all scientific and professional knowledge about filtering and retrieval disciplines. Failure to address the causal relationship between queries, documents, and performance can result in a weak science that makes few significant advances. Knowing the cause and extent of performance increases assists both the scholar and the searcher in understanding what query, what system, and what options will produce either improved or optimal system performance. In addition, knowledge of what causes performance to increase or decrease can direct empirical research toward those questions that currently cannot be answered analytically and must be studied through the examination of empirical data. For example, it will become obvious to the reader that the relationship between the frequencies of terms in the set of relevant documents and in the set of all documents plays a key role in determining the performance of filtering systems. This suggests that empirical studies of performance might best be directed at studying these frequencies and their accurate estimation, rather than, for example, comparing ranking formulae using experimental methods.

1.2 SYSTEMS AND EXPERIMENTS

When describing filtering and retrieval systems, we refer to books, articles, passages or text fragments that are viewed as single textual units as *documents*. Text, recorded expressions in natural or artificial languages, represents the bulk of useful material retrieved and filtered, although other forms of material such as still or moving images, as well as audio representations, are increasingly being combined with text in documents to be retrieved or filtered. Increasingly, media are being stored without the use of text to describe the content of the non-textual media. For example, a picture of a personal computer might be represented and searched for with a set of words, such as "computer, desktop, keyboard, CRT," or it might be searched for by determining the similarities that exist between the stored shapes and a query consisting of a shape or image, such as a picture of a computer.

The filtering of textual documents or media moving over a network and the retrieval of documents from a database are treated here as being very similar; the conceptual differences between them are relatively few. Document retrieval systems accept a query from a searcher and then evaluate all the documents in one or more databases, considering each document for presentation to the user (Salton & McGill, 1983; Van Rijsbergen, 1979). Some or all of the documents are ordered for display. After examining one or more documents, a user may then supply *relevance feedback*, judgements on the relevance of presented documents, which are then used by the system to better learn the characteristics of those documents that the user finds beneficial and the characteristics of those documents rejected by the user. The retrieval system may then present additional documents to the searcher after automatically adjusting the query or after the user has directly modified her or his query. The process continues until the user decides to stop the retrieval process.

Text filtering systems present only material expected to be useful to the end-user. Filters and agents applied to network traffic may operate in real-time, quickly making the decision to pass or reject a document in the fraction of a second before the next document arrives for consideration (Losee, 1989; Belkin & Croft, 1992). The decision to present a document to the user is made by a filtering system based on the document's characteristics. For example, a filter might allow only documents containing all three terms x, y, and z through the filter, excluding all other documents. For this reason, the choice of the characteristics used in making retrieval decisions is a critical determinant of the quality of the filtering. Filters are required to perform most of the functions performed by a comparable document retrieval system, except for the database management functions inherent in a retrieval system. Thus, we explicitly assume that most aspects of filtering system performance may be evaluated using the same measures as those applied to retrieval systems. Because the emphasis here is on the similarity in the decisions made by these two kinds of systems, we often refer below to filtering and document retrieval systems interchangeably.

A great deal of what has been learned about information retrieval has come from experimental research based on theoretically suggested retrieval and indexing models. The performance of these systems is usually studied by examining the qualities of both individual documents and groups of texts presented for possible use to a hypothetical searcher. Retrieval research usually assumes that a full experimental system may be viewed as a system of separate components. The quality and nature of these components directly determines the quality of the performance measures and the generalizability of these results to other environments.

By retrieval and filtering systems, we mean those parts of systems that contain and retrieve documents and that are situated outside human beings.

Retrieval systems, for purposes here, begin with a process that accepts a query and ends with the return of one or more documents. Thus, we do not address questions about why or how people develop information needs, solve problems, or think, leaving these important topics to cognitive psychologists. We also don't address why people decide to acquire one document instead of another, leaving this to those marketing researchers who have spent lifetimes studying consumer preferences, asking questions such as why people choose to purchase one candy bar instead of another, as well as how products should be arranged in a store to maximize the store's sales.

Most text retrieval and filtering experiments take as data a fixed set of queries, documents, and relevance judgements, collectively referred to as an *experimental database*. Many are composed of hundreds to thousands of documents, with the addition of databases for projects such as TREC (Text REtrieval Conferences) and MUC (Message Understanding Conferences) increasing the numbers of documents into the millions. Experimental databases are usually developed by extracting a subset of documents from a large commercial database, based on some broad criteria. For example, one might select documents having a particular term or expression (e.g., *information filtering*) to be included in the test database.

Representation and Access

There are many practical problems associated with analyzing the similarity between the desired characteristics expressed in a query and the characteristics of each of potentially millions of documents. Because of this complexity, manual and knowledge-intensive techniques have been used to support retrieval in paper-based collections. Using pre-computer techniques, libraries have long used classification systems (e.g., the Dewey Decimal system) to support subject browsing. This most commonly takes one of two forms. The first is a linear arrangement of documents by subject. Libraries usually arrange documents linearly, which may be understood conceptually as organizing books on a single, very long shelf. A second organizational form is as a set of clusters of documents, each cluster containing similar documents. Retrieval systems can determine the similarity between the characteristics of queries and the characteristics of each cluster. The similarity between a cluster and a query may be the distance between the query and the center of the cluster, or the distance between the query and the point in the cluster nearest the query or between the query and the point in the cluster furthest from the query. Those clusters most similar to the query are then subject to further analysis, such as by examining each document to determine which documents are most similar to the query.

If documents are to be retrieved, they need to be represented in a form that will capture the meaning of those features likely to be of interest to potential searchers, as well as to identify those terms likely to be useful when discriminating one document from other documents. Most document databases include enough text in a document's topical representation to act as a reasonably accurate surrogate for the full document. Failure to include document features that

might be of interest to searchers will result in the retrieval system being un-
able to discriminate between documents the user might find helpful and those
documents of little use.

Database systems are usually successful at capturing much of what a document is about by representing it with natural language terms occurring in
the document, or with human assigned terms or expressions capturing the
intellectual content of the document. Often, however, these terms do not represent adequately other, non-topical aspects of documents. For example, most
retrieval systems fail to represent fully many aspects of documents, such as
reading level, type font, style, cover art, or other factors that would help an
individual or a system identify a document as potentially useful. The author's
six year old daughter was able to accurately predict, after several years of casual
observation, which paperback books would be of interest to her father solely
based on the presence or absence of certain images, such as airplanes, on the
covers of the books her father reads. Most retrieval systems can't perform at
this level, or handle other subtle stylistic factors, usually because the needed
characteristics weren't entered into the document database when it was created,
or because the system is incapable of working with the larger structures and
sophisticated feature relationships that are needed to identify more complex
aspects of documents.

Representations are frequently topical, indicating the subject of the documents. Much of this information is carried by individual terms in the natural
language of the text. The weakness of always using natural language terms
taken directly from a title to represent a document is illustrated by considering
the American classic, *Gone with the Wind,* which describes life in the American
south in the second half of the nineteenth century. Searching for expressions
like *Civil War* or *south* will fail to retrieve this classic; instead, it will be retrieved by people interested in *wind.* Because of this kind of problem, human
reasoning has long been used in indexing and cataloging documents, so that
expressions like *civil war* might be assigned to books like *Gone with the Wind.*
Totally automated indexing and retrieval systems may eventually be able to
recognize complex relationships, such as the one between a query with the expression *civil war* and a document containing expressions like *confederacy* or
union soldier. Clearly there are reasons to think that the incorporation of additional processing when representing documents, using linguistic and "common"
knowledge, is likely to improve information filtering system performance.

How the information in natural language may best be extracted, or at least
most practically extracted, may be understood best by considering a scholarly
spectrum with two extremes. Scholars on one end might argue that all that
is present in language can be obtained from and understood by examining the
relationships between terms in statements, using statistical or logical means, as
well as the relationships between the terms and real-world events or phenomena.
Those at this end of the spectrum may believe that the full inclusion of linguistic
information into retrieval and filtering systems will come only with a major
increase in statistical processing power over what exists now in commercially

affordable retrieval systems, enabling massive numbers of relationships to be examined.

Those on the other end of the spectrum might argue that humans are unique beings, and that information and knowledge cannot be described in an objective fashion. Scientific methods will never be able to fully represent natural language without human intervention in the representation process, and even then, the representation will be less than perfect. There are philosophical arguments for both ends of the spectrum; perhaps one of the fruits of the increased interest in network filtering may be that many important questions about natural language phenomena will be answered sooner than they would have been answered otherwise.

In the methods described below, emphasis will be on the statistical relationships between text features. When knowledge and linguistic relationships are formalized as statistical relationships, they can be incorporated into more sophisticated models of filtering and retrieval. For example, the value of a term in representing the subject of a document may be estimated using quantitative or statistical techniques. If one considers terms that occur in this book that best represent what it is about, intuitively one rejects terms like *and, the, but, of,* and other very frequent terms as being subject bearing, while other less frequently occurring terms are more likely to be considered subject bearing. This suggests that term rarity is associated with the value of a term as an indexing term. Term rarity has been incorporated into the Inverse Document Frequency (IDF) weight that can be assigned to terms based on the frequency of the term in a document set. The IDF weight of a term is computed as $\log(N/n)$, where N represents the number of documents in the database and n represents the number of documents with the term (Sparck Jones, 1972; Croft & Harper, 1979). The IDF weight is higher for rare terms than for more common terms.

Consider the term *the* occurring in 999 of 1000 short documents. The IDF for this term, using natural logarithms (the *ln* key on many calculators), is $\ln(1000/999) = .001$. A less common term, like *retrieval*, occurring in only 2 of the 1000 documents, will result in an IDF of $\ln(1000/2) = 6.21$. The less frequently occurring term, *retrieval*, has a higher IDF weight than the more common term, *the*.

One potential drawback of using the IDF weight is that relatively rare and thus highly weighted terms may be so unusual that they may be poor indexing terms, as they represent topics which are unlikely to be searched. These include proper names and typographical errors.

Indexing may be unnecessary when the ranking algorithm to be used can formally incorporate term frequencies and between-term relationships. These systems can work from raw, un-indexed document representations, rather than from the output of indexing processes. These non-indexing retrieval procedures can learn what the user expects to find in relevant and non-relevant documents. Thus we can look for the the frequency of each term in a document, rather than looking for only human assigned indexed terms. Both non-indexed and indexed methods will continue to prove popular; technology and economics will determine which will dominate in the future.

Queries and Query Languages

The expression of an information need as a query presented to a retrieval system may be structured in several different ways, with features selected from several different sources (Iivonen & Sonnenwald, 1998). Probabilistic and vector retrieval models suggest that features in queries may be weighted individually, or groups of features may be given a group weight. Boolean queries specify the combinations of terms that must occur in documents if they are to be passed through a filter. Other models, such as fuzzy logic, may allow for additional combinations of weightings and connections, as well as provide different interpretations of expressed information needs.

Features in queries may be chosen so that they best represent what the searcher considers to be his or her information need, or queries may be predictive, expressing the searcher's guess as to which features are likely to occur in documents which would satisfy to some extent the information need that prompted the search. This is similar to the problem facing an adult addressing a child; should one express oneself as clearly and as precisely as possible in "adult" language, or should one use much simpler terms that the child might understand better but will not precisely represent what the adult wants to say? Users must decide to express the information need as they perceive it, or they may try to express it using terms and relationships that they think others would choose.

Queries may be small or large; they may contain only key terms, or a query may be a complex statement in a natural language. Users, especially untrained searchers, frequently select few query terms, resulting in lower levels of performance than would be obtained with larger numbers of discriminating terms. One study found that users often developed queries containing a few key terms or phrases; it was discovered later that additional relevant documents were retrieved when any of tens of other terms or phrases were used in a query (Blair & Maron, 1985). However, expressing an information need in a long natural language statement, such as *I want to see everything you have on the subject of X* will likely retrieve many documents containing non-topical terms like *subject, see,* and *everything.* Here, the addition of non-topical terms to the topical term *X* results in more system noise and decreased performance. Yet, precise small queries are also inadequate, and the optimal query size and the characteristics of an optimal query needs to be determined.

Another form of query is the *exemplar*: one or more relevant documents that can be used to represent what those relevant documents yet to be retrieved should contain. Those features occurring in a large proportion of relevant documents but in few non-relevant documents are good discriminators. These exemplars may be combined with other query forms to provide more information about the characteristics of relevant documents. Some existing systems do this in a fashion, accepting an initial query using a query language understood by the system, and accepting relevance feedback in the form of exemplars (Losee, 1988; Spink & Losee, 1996).

Often, a set of queries in an experimental database is formulated by a group of subject specialists. Usually these queries do not represent actual information needs of the query generator, but instead represent the types of queries they think are typical or that they might present to a retrieval system. These queries are thus not very representative of the domain of actual information needs or of how real needs are expressed.

Relevance Judgments

Relevance judgements represent what some consider to be the weakest aspect of the experimental database development process (Negoita, 1973; Bookstein, 1979; Swanson, 1986; Eisenberg, 1988; Saracevic, 1991; Schamber, 1994; Martin, 1995; Fidel & Crandall, 1997). The relevance of a document to a user may be interpreted in several ways. It may be understood as an objective, topical relationship between a query's subject and the subject matter of a document. The topical relationship may also be considered as a subjective match between an information need expressed in a query generated by a user, and the text. One user may find a particular document to be relevant to a query, while another user may consider the same document to be not relevant to their personal interests, even though the terms in the query may be the same for both the first and the second person. Subjective relevance is a relationship between the document and the internal state of the searcher at a particular point in time. This complex relationship (Schamber, Eisenberg, & Nilan, 1990; Barry, 1994) is multidimensional, with facets of relevance representing subjective phenomena, such as the needs or interests of the user, and what the user knows or has retrieved previously.

The user may find useful relevant material that was not directly mentioned in the query. For example, a user seeking information that is only available in another language may need access to a dictionary before being able to use the material. The subject of the material was mentioned in the query but the dictionary was not. Similarly, a user accessing a traditional automated system in the workplace may need additional guidance if they are unfamiliar with the operation of the system, the functions provided by the system, or the interpretation of the results produced by the system. These use-related problems are retrieval problems.

Experiments conducted using objective, topical relevance judgements may be generalizable to other similar environments. Results using subjective, personal relevance judgements may be generalized only to the extent that those who generated the personal relevance judgements reflect the commonalities and diversities in the population. When developing an experimental database, documents are presented to a subject expert for consideration of topical relevance. Commonly this set of documents is retrieved using a number of methods, the goal of this initial search being to retrieve as many documents as possible that might be relevant. With larger databases, it is clear that subject specialists cannot exhaustively examine the large number of documents for each query to determine their possible topical relationships with the query. Hence, the larger (and more realistic) the database, the less likely it is that the relevance judgements are accurate or complete.

In some cases, relevance values differ across individuals and the capturing and use of these differences and the similarities present is one of the functions of the system. Collaborative filtering systems, also known as recommender systems (Resnick & Varian, 1997), combine judgements from groups of people and provide recommendations about various topics, including, for example, movies and restaurants. In these systems, it is explicitly recognized that there will be different judgements based on personal tastes, but that if the reader and the author both liked books B, C, and D but neither liked books F or G, then if the author liked book H, there is a good chance that the reader will like book H too. Collaborative filtering systems provide these "personalized" recommendations based on judgements by people with tastes similar to those of the searcher.

For purposes below, the set of relevant documents is defined as all and only those documents the user desires to retrieve. Doing so places a *pragmatic definition* on relevance and avoids some of the problems associated with other definitions of relevance. It is far easier to learn the characteristics of what the user does (e.g. label a document as *relevant* or *non-relevant*) than it is to learn what the user is thinking or needs, or learn an abstract, possibly objective relationship between documents and topics that exists in another domain.

Ranking Algorithms

The relative effectiveness of different ranking procedures is a popular research topic for retrieval scholars, and a wide range of ordering procedures continue to be used commercially and studied academically. A popular ranking algorithm used in early computerized retrieval systems assumed information needs were expressed as Boolean statements, queries containing terms with operators *and*, *or*, and *not*. A Boolean query can be applied to the features of a particular document, with those documents that result in the value of the query being *true* being presented to the user. For example, a query on filtering performance might be *filter and performance* or, more elaborately, *(filter or filtering) and (performance or measure or measurement or measuring)*.

A prominent model relating queries and documents is the vector model, which suggests that both queries and documents may be viewed as lists or vectors of terms, and the degree of similarity between a query and a document is the cosine of the angle between the two vectors. A large angle between them indicates a high degree of dissimilarity, while a relatively small angle and a corresponding large cosine value represents a high degree of similarity.

Derived from decision theoretic considerations, probabilistic models of the decision to retrieve or reject documents can be used to rank documents. Given the expected utility of a document to the user and the probability of it having a particular relevance status, a weight can be computed that can be used to rank the set of documents. Probabilistic models of retrieval are examined below more than other models because the expected performance of probabilistic systems can more easily be determined and interpreted than it can for other retrieval models.

There is some evidence that combining different ranking methods to yield a new method can improve performance, possibly through capturing the best of the different methods (Losee, 1988; Hull, Pedersen, & Schutze, 1996).

System Research

Computerized retrieval research has a history going back over half a century. Early information retrieval system research was based on bibliographic data recorded on computer cards that were sent to different output bins on a punched card sorter. As technology advanced, the data moved from punched cards to magnetic tapes to magnetic discs to CD-ROMs. Earlier retrieval systems had little mathematical processing power available, and thus relied on simple query and document matching techniques.

As processing speeds increased, more complex experiments could be conducted. Tests began on different kinds of indexing systems, ranking procedures, and query languages. Representation studies included questions about whether *uncontrolled* vocabulary, natural language terms supplied by the user, resulted in better performance than with *controlled* vocabulary, a set of standard terms used for any of a number of synonyms (e.g. *cat* used for *cat, kitten, kitty,* and *feline*). The results of these experiments varied, with uncontrolled vocabulary often performing as well as manually assigned controlled vocabulary. This suggests that using uncontrolled vocabularies may be a viable commercial option, given the high cost of paying a human to assign controlled vocabulary terms to documents, and the much lower cost of having a computer assign index terms.

The other main research thrust was into query languages and weighting systems. Studies showed that models based on term weighting performed as well as Boolean queries, and that treating individual terms as independent worked reasonably well. This suggested that terms could be extracted from natural language queries and then used satisfactorily for query terms, removing the need for Boolean queries and the searcher's understanding of Boolean logic. This opened the door for end-users to routinely do their own searching, decreasing the need for search experts to act as intermediaries. As increased processing power and memory have become available, it has become possible to use more complex mathematical techniques in real-time, such as those incorporating sophisticated statistical relationships between terms to improve performance.

More recently, research and the resulting commercial systems have focused on handling the rapid growth in the number of documents being distributed over international networks. This has required the development of filtering and retrieval techniques that can work with millions of documents and large numbers of queries at a high rate of speed. Some systems support sets of documents spread geographically over large areas. Many of the current systems index and point to millions of World Wide Web pages throughout the Internet and can handle millions of queries a day.

Experimental and commercial systems are evaluated using any of a number of performance criteria (Van Rijsbergen, 1979; Salton & McGill, 1983; Losee, 1990). Many systems are evaluated by looking at the quality of the set of documents retrieved as the search progresses toward the retrieval of all the useful documents. Comparing systems using these measures may involve making value judgements about the relative importance of different parts of the search. To avoid this problem, single-number measures of filtering performance have been proposed which provide a single performance figure for each retrieval technique. Ideally, one can then select the retrieval technique with the best performance number.

1.3 ANALYTIC METHODS

An analytic theory of text filtering and retrieval performance provides a set of mathematical statements that describe the relationships between system variables and the performance obtained with the system. A theory of the operation of text filtering and retrieval systems should:

- *describe* current performance,

- *predict* future performance, and

- *explain* performance.

For a theory to be worth considering seriously, it must describe the performance that occurs. In the following chapters, we will provide examples and show that the performance obtained with the example data is the same as that described by the analytic model.

Better than this is a theory that can predict accurately the level of performance for queries, documents, and relevance judgements that have never been found together and which, when studied, have the results predicted by the theory. Using the analytic model, for example, one can examine the effects of different assumptions about various components in the system. More importantly, the model can make predictions about systems consistent with assumptions that have never been brought together in a particular environment. This predictive capability is valuable for both the scholar and the manager since both desire to determine which model or system will be superior in their environment prior to purchasing and installing the system.

A third level of theoretical adequacy is the ability of the theory to provide an explanation of system performance, a general statement of the relationships between variables and events. An explanation of text system performance is a set of statements about the interrelationship among factors such as term frequencies, relationships between terms, user relevance judgements, and the grammatical assignments made to variously sized text fragments. The development of methods that can be used to describe, predict and explain text filtering and retrieval performance is the purpose of this book.

As an example of the successful modeling of system performance in another field of study, one can turn to queueing theory, which can be used to describe, predict, and explain the operation of some aspects of computer system performance. While queueing theory can often accurately describe system performance, it also can make inaccurate predictions. This is not usually because of a weakness in the queueing theory itself, but because of the inappropriate or uninformed application of the theory to situations where the assumptions of the model are not met. Queueing theoretic models make statistical assumptions and are expected to be accurate only when those assumptions are met or closely approximated. The successes (and failures) of queueing theory serve as a guide to the strengths and weaknesses of the analytic models developed below that relate queries, documents, relevance, and term relationships provided by natural language. *When the model's assumptions are met, the tool is invaluable in predicting and understanding performance; when the assumptions are not met, the tool must be used with care.*

The quality of the results produced using a model can be determined in part by how well the assumptions of the model are met. This may be determined by comparing distributional assumptions made by the model to the actual data. Statistical tests such as the chi-square test can be used to determine whether a dataset differs significantly from an assumed distribution. In other cases, one can compare how the model performs with a live data set and then compare those results to how it performs with simulated data that meet the statistical assumptions made by the model (Losee et al., 1986). Using *sensitivity analysis*, one can apply different reasonable models to determine the degree to which assuming different models results in a decrease in performance. One then learns the robustness of the model and its assumptions. In the case of a rather robust model, simplifying assumptions may be made with little loss in predictive ability.

When Analytic Studies are Superior to Experimental Studies

Analytic studies often provide different and more useful results than experimental studies. One reason is that analytic results are independent of a particular experimental database, being dependent only on the values assigned to model parameters. The results produced can be generalized to any environment with similarly valued parameters. The analytic model allows us to focus on these important features and find the characteristics of successful systems.

There are many other circumstances where experimental methods cannot state clearly that one ranking method is better than another ranking method, if the difference between the methods is small. When there are small performance differences between methods, the analytic model can discern that there is, in fact, an expected difference, and supply the exact magnitude of the expected difference. Experimental results showing that two different methods perform slightly differently might be detecting a true difference between the two methods, or the observed difference might be a random fluctuation in results. A method that is superior might appear experimentally to be inferior, given the particular database being used. These problems are avoided by analytic studies.

Database Size

A recurring question when studying retrieval and filtering systems is the extent to which experimental studies using small numbers of documents can be generalized to larger, more realistic databases of hundreds to hundreds of millions of documents. Experiments using small databases may produce results that will fail to be statistically different from each other but the results would be significant if the test were conducted using a multi-million document database. In addition, it can be shown that "tests on collections of a few hundred documents are not sufficiently sensitive to permit even rough estimates of performance on very large collections of hundreds of thousands of documents" (Swanson, 1977).

The relationship between database size and the level of performance experienced has been suggested to be a non-linear relationship. Blair and Maron (1985) note that "the amount of search effort required to obtain the same recall level increases as the database increases, often at a faster rate than the increase in database size." The difference in performance due to size differences is difficult to study effectively. This is due, in part, to the wide range of document densities and database sizes found in realistic databases and text files moving through networks. In some sets of documents, there are relatively few relevant documents, and in other sets, large percentages of the documents are relevant. Some sets of documents are very homogeneous, with the non-relevant documents having many features in common with the relevant documents, while other databases have very different documents, with relevant documents being easily separated from the non-relevant documents, based on their features. Studying these options is very time consuming, both when developing the experiment and when conducting the test itself. These studies are also very expensive to conduct, with the costs increasing in proportion to the degree of realism. On the other hand, it is easy and inexpensive to analytically study these variations in database size and structure, and performance variations are easily compared.

Model Verification

Analytic models provide a theoretical basis for describing performance results. Support for such a theory is provided by experimental evidence that is consistent with what is predicted by the theory. For example, while we present data here that show this level of consistency in simple cases, additional support would allow us to determine whether it is equally able to describe large system performance correctly, scaling to larger systems and producing results that are consistent with the assumptions of the analytic model. One of the advantages of analytic models, such as those suggested below, is that they would appear to scale effortlessly, suggesting that if models are consistent with small data sets they are likely to be consistent with larger systems that meet the assumptions of the model.

Models are based on different kinds of scientific theories. One can propose an empirical relationship between energy, mass, and the speed of light ($e = mc^2$), or a relationship between speed, time, and the distance covered. These empirical relationships are based on what was initially a hypothesis, a guess based on data whose analysis suggested the relationship. There are other models that are also empirically testable but are derivations from other accepted laws or methods. The analytic model developed below is more similar to a derivative form of theory than it is to a hypothesis-based theory.

Models such as these may be validated by comparing analytically predicted results with empirical results produced by experiments and simulations. One problem in verifying the models is determining whether experimental results obtained are accurate. Computer programs that have been written to implement experimental systems and measure the results may themselves have "bugs" in them. If there is a disagreement between theoretical prediction and experimental results, which is to be believed? While the theory could be wrong, it is also quite possible that the experimental system does not produce documents in the order or the way that the users and designers of the system believe it does. While finding *correct* contrary empirical examples would indicate that there is a flaw in the model, an easier and better means of falsifying a derivative model is to attempt to understand the model and then argue that it is flawed, possibly providing a corrected model. The reader is strongly encouraged to consider how the work below could be improved upon; there is a lot of room in the analytic study of filtering and retrieval performance for new, expanded, and more elaborate models!

Experimental Results

When studying a retrieval system using two different experimental methods, one might wish to assert that one method is better than the other and that the difference in results is statistically significant. This is most commonly done by taking a set of performance measurements for several different queries using each method. The two sets of measurements are then compared and, ideally, the one that has been proposed to have better performance has significantly better performance than the method to which it is compared.

Determining the statistical significance of the performance variations obtained with different filtering methods is difficult. One common method uses the statistical t-test. The t-test compares the differences between two performance means to determine whether they are likely to come from the same population (no statistically significant difference) or are from different populations (a statistically significant difference). Other significance tests and approaches can be used, such as the sign test, that makes make fewer parametric assumptions than the t-test, or studying whether retrieval performance is above one of the better retrieval performance levels that would be generated randomly (Shaw, Burgin, & Howell, 1997). These tests can be used to argue that two

ranking methods are truly different and that one is significantly better than
the other. Yet, even given a significance test such as one of these, we still don't
know that one method is better than another and, if they are believed to be
statistically different, we don't know *why*. Certainty can be obtained *only* from
an analytic model.

Experiments and data collection still have a major role in the study of information retrieval. Data collection is needed to learn the parameters of different
environments. Experiments allow one to validate models and to further study
real-world phenomena that cannot be modeled accurately.

1.4   QUESTIONS NEEDING ANSWERS

Numerous questions about filtering and retrieval systems need to be answered,
and the answers understood, if scholars and professionals are to develop and
use effective systems. Most of these questions can be addressed, fully or in
part, by using analytic methods.

One area that needs to be understood is how a user presents his or her information need or interests to a system. There are two ways to address this
area. The first is through the study of what is best for humans, that is, what
is the most natural or least error prone means of expressing such a need. This
question, which might be most properly considered a question about cogni-
tion and learning rather than a question about filtering and retrieval, is best
answered experimentally, since there are no widely accepted models of how
humans express information needs that are capable of accurately predicting
specific actions and preferences. Questions about the efficiency of means for
expressing the information need to the system may be addressed and experienced analytically. These questions include:

- What can a system extract from natural language statements of information need, given specific models of language, and including such features or
  information sources as grammars or dictionaries?

- What is the performance associated with specific natural or artificial query
  languages? Are some better than others in specific circumstances?

- How does the length of a query and the depth of the expressed information
  need affect performance?

- How do different techniques for estimating parameters affect performance?

As with queries, databases may vary in a number of ways. Answers to the
following will result in increased understanding of text filtering and retrieval
and improved performance:

- What is the scale effect as documents are added to or deleted from the
  database and the size of the database changes?

- What will be the performance effect of changing the relative number of the
  relevant and non-relevant documents independent of each other?

- How does the failure of documents to meet statistical assumptions made by models affect filtering and retrieval performance?

Ranking algorithms are used to order the documents, and since the ranking directly affects the performance, understanding the direct relationship between the ranking algorithm and the performance figures is critical to the understanding of filtering and retrieval:

- What are the strengths and weakness of specific ranking methods?

- What are the relative strengths and weaknesses of probabilistic versus logical methods?

1.5   THE STRUCTURE OF THIS BOOK

Many of these problems will be addressed as an understanding of retrieval and filtering is developed. The next chapter will present some of the fundamentals of probability theory necessary to understand basic retrieval models and the effects produced by specific query types, documents characteristics, and assumptions of text filtering and retrieval models. Information retrieval and filtering models will be discussed next, followed by a discussion of performance measures. While these measures will include traditional experimental measures, we will also examine analytic measures, including the intrinsic degree of optimality of different formulae used in assigning weights to documents for ranking. After this, Chapter 6 will present methods for predicting performance for filtering and retrieval systems when a single term is used for the query or filter. While a query with a single term may seem simplistic, it serves as the basis for the most simple models of text filtering, and understanding these simple models will allow us to advance to more sophisticated models in later chapters. After presenting single terms, there is a chapter each on multivariate probabilistic methods and on the application of these techniques to multi-term filtering models. Then we examine techniques for understanding the effects of logical statements and the relationships between terms on retrieval performance, including the knowledge supplied by natural language statements.

2 QUANTITATIVE REASONING FOR FILTERING

Mathematics, rightly viewed, possesses not only truth, but supreme beauty—a beauty cold and austere....

— Bertrand Russell, *Study of Mathematics*

2.1 INTRODUCTION

How does one make inferences about the future, and how does one reason about the present? Reasoning requires that one relate, in some fashion, one event with another. Probability theory and predicate logic are the two most popular systems used to describe these relationships between states of nature. Probability theory can provide us with tools to compute the *expected value* of performance, the best guess as to the results that will occur.

A probability represents the chance of a random event occurring. Denoting the chance of event x occurring as $\Pr(x)$, a probability may be interpreted in either of two ways. The *long run* or *frequentist* interpretation of probability suggests that the probability of an event is the limit of the proportion of the event's occurrence having a characteristic, as the number of observations approaches infinity. Thus, we may say that a fair coin has the probability of *heads* of .5 because, if one flipped a coin a very large number of times, very close to 50% of the coin tosses would produce *heads*. The subjectivist model of probability suggests that the probability of .5 represents our belief or best guess that an individual coin will land as *heads.* One argument for this subjectivist view of probabilities is based on the difficulty with estimating the chance of a one time event that is to occur in the future. For example, the probability that one will die tomorrow is difficult to compute in terms of long-run probabilities; this will be a one time event.

A variable such as height, weight, or whether a document contains a particular term is a *random variable,* with the value varying across individual events. A random variable is usually denoted with an uppercase variable (for example, X) and a specific instance of the variable is denoted with a lowercase variable, or the actual value is indicated. Depending on whether a specific value or a more abstract value is intended for the variable, the variable's value may be denoted as $X = 2$, $Sex = f$, or $A = a_i$. The probability that a random variable X takes on a particular value x may be denoted as $\Pr(X = x)$ or $\Pr(x)$.

A random variable X is distributed over a range of values. The *probability density* function provides the probability at each point in its range of values. All the probabilities for the distribution of random variable $X = x_1, x_2, \ldots, x_n$ sum to 1, that is,

$$\sum_{i=1}^{n} \Pr(x_i) = 1.$$

Distributions of random variables may be *discrete* distributions or *continuous* distributions. Discrete distributions are found where the values with associated probabilities are discrete units, such as integer (whole number) values. For example, the random variable *number of children in a family* has values for no children, one child, two children, three children, etc. These integer values are commonly found in retrieval and filtering applications where the number of documents or terms or users is always an integer. Continuous variables are found with quantities that vary smoothly, such as time or mass or width. For example, the time that has elapsed since you started reading this book is usually treated as a continuous variable. There are an infinite number of possible times between any two other times; time, if measured infinitely precisely, is a continuous variable.

The sum of the probabilities from the lowest possible values to i is the *cumulative distribution function* or just the *distribution* function. The distribution function $C_i(X) = \Pr(X \leq i)$ is computed as

$$C_i(X) = \sum_{j=-\infty}^{i} \Pr(x_j)$$

for discrete distributions and

$$C_i(X) = \int_{-\infty}^{i} \Pr(x_j) \, dj$$

for continuous distributions.

A probability may be treated as a *point* probability, a single value representing the probability that an event occurs, or it may be treated as a distribution, representing a set of probabilities associated with the set of values an event may have as the true state of nature. For example, one might wish to estimate the point probability that an event occurs as 1/4. One might instead estimate the set of different probabilities describing the underlying parameter as a distribution that peaks near 1/4. An individual might have their own personal belief that it is most likely that the probability of the event occurring is 1/4 but that other probabilities are possible, with the distribution describing the relative degree of confidence in the different probabilities. These are probabilities about probabilities.

EXERCISE 2.1. [*Easy*] Which of the following are best treated as continuous random variables and which as discrete random variables: (1) a person's weight, (2) the number of answers on a multiple choice test that are correct, (3) the age of an individual, measured in whole years, and (4) the time it takes to read this page.

Central Tendency

The *mean* of a set of values or of a probabilistic distribution is a measure of central tendency, a value that can represent the "middle" of the set of data. The mean is computed as

$$E(X) = \frac{\sum_{i=1}^{n} x_i}{n} \tag{2.1}$$

where x_i is the ith value for x, and n is the number of items. The $E(X)$ notation denotes what we might expect of the random variable X, the *expected value*, and is used to represent the estimated mean. The arithmetic mean is such that the following expected value is minimized when μ is the mean: $E((X - \mu)^2)$. The mean is the center of mass of the distribution describing the data.

The median is such that the value $E(abs(X - \mu'))$ is minimized when μ' is the median for the data set described by random variable X. The median will be such that half of the distribution is above it and half is below it.

EXERCISE 2.2. [*Easy*] What is the mean of the set of numbers: 4, 5, 6, 7, 8, 9, and 17? What is the median? How does one of these represent the "center" of the random variable better than the other?

Table 2.1. Sample data for computing joint probabilities. Using this data, the number of males who prefer watching TV to reading books is 4.

	Preference:		
	TV	Books	
female	1	2	$1 + 2 = 3$
male	4	3	$4 + 3 = 7$
	$1 + 4 = 5$	$2 + 3 = 5$	$3 + 7 = 10$ or $5 + 5 = 10.$

Variance

Variance is a measure of the spread of data, the dispersion of the data around the mean. The variance is computed as

$$V(X) = \frac{\sum_{i=1}^{n} (x_i - \mu)^2}{n}, \tag{2.2}$$

where μ denotes the mean and $V(X)$ denotes the variance of the random variable X. The variance is the average squared difference between the individual values and the mean.

The standard deviation, σ, the square root of the variance, is often more convenient to use than the variance, σ^2. The standard deviation is in the same units as the values from which it is estimated and is thus easier to work with. For example, the standard deviation of the heights for human beings may be measured in inches or meters, while the variance would be in inches squared or meters squared, units more difficult to work with than simple inches or meters.

EXERCISE 2.3. [*Easy*] What is the variance of the set of numbers: 4, 5, 6, 7, 8, 9, and 10?

EXERCISE 2.4. [*Easy*] What is the standard deviation for the preceding exercise?

Joint Probabilities

The data in Table 2.1 show that the probability a randomly sampled person in this group will be *female* is $3/10 = .3$. This is denoted as $\Pr(female) = 3/10$ and is computed as the row total for females divided by the grand total in the bottom right hand corner of the table. The probability of a simple unconditioned event, representing one of the rows or columns in the table, is a *marginal* probability. We can compute the probability that someone is *female* and prefers *books* as the number in the cell for the *books* column and the row *female*, 2, divided by the total number of people, 10, producing a probability of .2. This is the *joint* probability of being female *and* preferring books. Joint probabilities are denoted with a comma separating the variables to be joined: $\Pr(female, books)$.

The joint probability of x and y occurring together is related to the conditional probability $\Pr(x|y)$, read as the probability of x, *given* y, as

$$\Pr(x, y) = \Pr(x|y) \Pr(y).\qquad(2.3)$$

Thus, the probability that x and y will occur together equals the unconditional chance that y occurs multiplied by the chance that x occurs given that y is true. Using the data from above, the probability that an individual is female given that they prefer books is the percent of females in the books column, 2 out of 5, or .4. If we multiply $\Pr(female|books)$ times $\Pr(books) = .5$, one arrives at $\Pr(female, books) = .4 \times .5 = .2$.

It is always the case that

$$\Pr(x|y) \geq \Pr(x, y).$$

If we consider $x = female$ and $y = books$, $\Pr(x|y)$ is the cell value 2 divided by the number of preferring books, 5, or $2/5$, while $\Pr(x, y)$ is the intersection of the two, 2, divided by the total number, 10, or $2/10$.

In the case where x and y are statistically independent, meaning that knowing the value of the variable x provides no information about the value of the variable y,

$$\Pr(x, y) = \Pr(x) \Pr(y)\qquad(2.4)$$

EXERCISE 2.5. [*Easy*] Using the data in Table 2.1, what is the joint probability that an individual is male and prefers TV?

EXERCISE 2.6. [*Easy*] Using the data in Table 2.1, what is the conditional probability that an individual is male, given that they prefer TV?

2.2 DISTRIBUTIONS OF RANDOM VARIABLES

When using probability theory, one often makes use of one of several standard distributions that describe the probabilities that random variables have a given value from within the distribution's range. Data may be viewed as produced by a specific random *process*. For example, tossing a coin is a random process, with different flips of a coin possibly producing different values (heads or tails). While in the real-world coins are not perfectly symmetrical, one can gain an understanding of what is occurring and with great accuracy over time what will occur by assuming that this is a binary process with the probability of *heads* or *tails* of .5. Probability distributions often describe exactly the probabilities of particular events, but in many cases simple random processes are used successfully as approximations for reality. These simpler systems provide explanations that may be easier to understand, or may capture an underlying phenomenon of importance, while missing some less important processes. While it is always desirable to be exact, the choice is often made to accept small errors at the cost of greatly simplifying a model.

Human activities are never described perfectly by any common statistical distribution, yet such descriptions provide insight into many (but not all) of the processes being observed. Most models of text filtering and retrieval make assumptions about the distributions of term occurrences, relevance categories in which documents occur, as well as more complex assumptions about other relationships. We will often accept these assumptions, but will need to consider the tradeoffs associated with accepting such assumptions.

Binary Distribution

The Bernoulli distribution, most commonly called the binary distribution, describes the probability of a binary or dichotomous two-valued event, given the relationship between the probability of a value occurring and the value of the binary feature. The binary features are usually parameterized so that they have either the value 0 or 1. Computed as

$$\mathcal{B}(x; p) = p^x (1 - p)^{1-x}, \tag{2.5}$$

the binary distribution yields the probability that x will be a 1 or 0 given the value of p. The parameter p may be interpreted as the probability that x will be 1 or as the percent of occurrences that are 1. In the case where $x = 1$, Equation 2.5 reduces to p^1, while when $x = 0$, it reduces to $(1 - p)^{1-0} = 1 - p$. Characteristics of the distribution include the mean, $E(X) = p$, and the variance, $V(X) = p(1 - p)$.

Computing the probability of a binary variable may seem to be a matter of common sense. Clearly, if there are only two options, and the probability of one is known, the probability of the other is simply one minus the probability of the first. If 30% of a school's students are "in-state" students, then 70% of the students are "out-of-state" students; if 55% of a local college's students are female, then 45% are male. The Bernoulli distribution formalizes this knowledge and rises above this common sense procedure; the binary distribution can be manipulated formally, allowing us to further study the nature of phenomenon in which the binary process occurs.

Example.   Assume that there are 100 documents, 25 of which are relevant (represented by a relevance value of 1). The parameter p is thus $1/4$. We can compute the probability a document is relevant ($r = 1$) or not relevant ($r = 0$) as $p^r (1 - p)^{1-r}$. The probability that a document is relevant, that is, $r = 1$, is $(1/4)^1 (1 - 1/4)^{1-1} = 1/4$. The probability that a document is not relevant ($r = 0$) is $(1/4)^0 (1 - 1/4)^{1-0} = 3/4$.

EXERCISE 2.7. [*Easy*] Random variable D is produced by a Bernoulli process, with $p = 1/3$. What is $\Pr(D = 0)$?

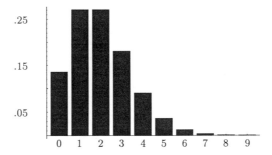

Figure 2.1.   A histogram for the Poisson distribution with a λ of 2

EXERCISE 2.8. [*Moderate*] Consider two binary distributions, with the first having parameter p and the second q. What is the ratio $\Pr(d; p)/\Pr(d; q)$?

Poisson Distribution

The Poisson distribution (named after the Frenchman Simeon D. Poisson, and only coincidentally the French word for *fish*) computes the probability of having x events, given that the events occur with an average frequency of λ occurring per unit of time. In addition, these randomly occurring events are *memoryless*, that is, the time when one event occurs does not depend in any way on when its predecessor occurred.

The probability of x events, given the average frequency of λ, is

$$P(x; \lambda) = \frac{e^{-\lambda} \lambda^x}{x!},\qquad(2.6)$$

where x factorial,

$$x! = x(x-1)(x-2)\cdots 1,$$

represents the product of all the numbers from x down to 1, and $0! = 1$. That $0! = 1$ may be understood by noting that $(n-1)! = n!/n$ and thus $0! = 1!/1 = 1$. One will often see used in place of the factorial function the gamma function, $\Gamma(x) = (x-1)!$ This is not to be confused with the gamma distribution. The gamma function may be treated as a continuous form of the factorial function. In Equation 2.6, the value e is a mathematical constant equal to 2.718. Both the mean and the variance of the Poisson distribution is λ.

Assume that terms are produced by authors writing documents about a selected topic in such a way that authors write with a certain average frequency of occurrence λ of a term in these documents. In addition, we assume that the term production process is *memoryless*. While this is obviously not met, it may be a satisfactory in many applications as an approximation of term occurrences in larger texts. If $\lambda = 2$, we arrive at the probabilities for the term occurrences:

no time, one time, two times, and three times (from the distribution shown graphically in Figure 2.1):

x	$\Pr(x; \lambda)$	$C_x(X)$
0	.1353	.1353
1	.2707	.4060
2	.2707	.6767
3	.1804	.8571

EXERCISE 2.9. [*Easy*] Assuming that terms are Poisson distributed, what is the the probability a term will have a frequency of less than 2 if the average term frequency λ is 1?

EXERCISE 2.10. [*Moderate*] As in the preceding example, assume that terms are Poisson distributed with $\lambda = 1$. What p value for the binary distribution will produce the same probability that a term will have frequency 0 as these Poisson distributed terms?

Negative Binomial Distribution

The negative binomial distribution, also called the Pascal distribution when the model variables are limited to integers, is the probability that there will be a total of r failures given that there are n successes, with p being the probability of the binary event being a success:

$$\mathcal{NB}(n|r, p) = \binom{n + r - 1}{r} p^n (1 - p)^r. \tag{2.7}$$

The binomial coefficient,

$$\binom{n}{x} = \frac{n!}{x!(n - x)!}$$

is found frequently in combinatorics, where it represents the number of distinct ways one can select from n events something containing x events. Note that the choice of the terms *success* and *failure* are arbitrary and they can be exchanged if this leads to a greater degree of understanding and is done consistently.

The mean for the negative binomial distribution is

$$E\left(\mathcal{NB}(n|r, p)\right) = \frac{n(1 - p)}{p}, \tag{2.8}$$

while the variance is

$$V\left(\mathcal{NB}(n|r, p)\right) = \frac{n(1 - p)}{p^2}. \tag{2.9}$$

Any Poisson distribution may be easily approximated as a negative binomial distribution with proper parameterization. The Poisson distribution has its mean equivalent to its variance. In situations where data appears Poisson-like but has a variance greater than or less than the mean, the negative binomial distribution associated with the Poisson distribution may have its variance modified to fit the dataset, providing a more accurate description.

Example.   Assume that a document is produced such that it is relevant to a topic ($p = 1/100$) or is not relevant. If there are 100 non-relevant documents produced ($p = 99/100$), the negative binomial distribution describes the probability of how many relevant documents will be produced:

x	$\Pr(x; \lambda)$	$C_x(X)$
0	.3660	.3660
1	.3660	.7320
2	.1848	.9169
3	.0628	.9797
4	.0162	.9959

Normal Distribution

Most distributions used to describe document features are discrete. Features may be continuous variables in some circumstances, such as when they contain a value describing a physical event or state, such as one's weight or the rate of use for a retrieval system.

The normal distribution is frequently used to model bell shaped continuous distributions. Describing the probability that a random variable has a given value, where the random variable may take on any value in the range from $-\infty$ to $+\infty$, the normal density function is

$$\mathcal{N}(x; \mu, \sigma) = \frac{1}{\sigma\sqrt{2\pi}} \exp\left(\frac{-(x - \mu)^2}{2\sigma^2}\right), \tag{2.10}$$

where μ is the mean and σ the standard deviation. It is sometimes helpful to remember that about 68% of all data will lie within plus or minus 1 standard deviation of the mean, while about 95% will lie within plus or minus two standard deviations.

In most cases, the exact assumptions of the normal distribution aren't met, but they may be close enough to the true state of nature to allow for the normal distribution to provide an approximation of the actual distribution of the data. For example, an assumption of the normal model that is widely violated is that the data extends to the values of $-\infty$ and $+\infty$. Many tests, however, are treated as having scores that are normally distributed. For example, most adult IQ tests have $\sigma = 15$ and $\mu = 100$, while the Graduate Record Exams (GRE) and Scholastic Aptitude Tests (SAT) have $\mu = 500$ and $\sigma = 100$. While having published standard deviations and means allows one to easily compute what percent of test-takers have a score within a certain range, this scoring violates the assumption that scores can extend beyond the end of the practical test limits. Nobody can score below 0 on the most common adult IQ test (WAIS-R) and nobody can score below 200 on either the SAT or the GRE test, but the normal distribution predicts there will be some individuals with scores below these cutoffs.

The normal distribution may also be used as an approximation for the Poisson distribution when the average for the Poisson distribution is large. Given the superior mathematical tractability of the normal distribution (especially when computing multivariate probabilities), it will often be helpful to use the normal distribution to model term frequencies rather than the Poisson distribution. This is clearly inappropriate when dealing with very small text fragments, such as titles or abstracts, but the normal assumption becomes increasingly useful when full-text is used, where the average term frequencies become much larger than is found in just abstracts or titles.

Other types of distributions may also prove useful when modeling. Probably the most interesting is the maximum entropy distribution, which will be examined more fully when discussing multivariate probabilities.

EXERCISE 2.11. [*Easy*] What distribution would be appropriate for computing (1) the probability one is under 5 foot in height (2) the probability 3 people might arrive at a stop sign during a given minute (3) the probability that a telemarketer will have to make exactly 100 phone calls while trying to get 4 sales, with the probability of making a sale on a given call being 1/20?

2.3 INFERENCE AND CONDITIONAL PROBABILITIES

While descriptive statistics such as the mean and variance are widely used, inferential statistics allow one to learn about a population and to make claims about the population given a sample of data. Being able to estimate the characteristics of certain categories of documents or terms from a few instances leads to improved text filtering.

Bayes' Rule

As well as being able to estimate probabilities based on the long term frequencies of events, one can move from effects back toward causes. The Reverend Thomas Bayes, an eighteenth century Presbyterian minister, proposed a rule now known by his name that can be used for such inferential tasks. Bayes' rule formally provides a means for computing a conditional probability given other, related probabilities. Applying this reasoning, using methods suggested by Bayes and Laplace, one can move probabilistically from effect to cause or from cause to effect.

Bayes' rule may be developed simply, beginning with the assumption that

$$\Pr(x, y) = \Pr(y, x).$$

Expressed less formally, the probability that the author has brown hair and a beard is the same as the probability that the author has a beard and brown hair. By substituting $\Pr(a|b)\Pr(b)$ for $\Pr(a,b)$ (Equation 2.3), we have

$$\Pr(x|y)\Pr(y) = \Pr(y|x)\Pr(x).$$

By dividing both sides by $\Pr(y)$, one arrives at the following definition of a conditional probability:

$$\Pr(x|y) = \frac{\Pr(y|x)\Pr(x)}{\Pr(y)}. \qquad (2.11)$$

Using Bayes' rule, one can compute conditional probabilities in terms of other probabilities. While no serious mathematician would dispute this derivation, certain applications and interpretations of "Bayesianism" are questioned by some scholars, particularly those who oppose the treatment of probability as inherently subjective and who instead support the frequentist approach.

Evidence supporting a Hypothesis

Bayes' rule may be thought of as describing evidence (e) and the relative degree of support it provides for a hypothesis (h):

$$\Pr(h|e) = \frac{\Pr(e|h)\Pr(h)}{\Pr(e)}. \qquad (2.12)$$

Consider here the initial hypothesis (unconditional probability) that someone is female, and consider what the evidence that someone prefers books adds to our knowledge about whether someone is female. We may begin with the *prior* probability, $\Pr(h)$, representing our prior knowledge about the hypothesis. Using the data in Table 2.1, the probability that someone is female is $3/10$. By incorporating the prior probability that someone is female with the likelihood, $\Pr(e|h)$, and normalizing by dividing by the marginal $\Pr(e)$, we can compute the *posterior* probability $\Pr(h|e)$, the probability that the hypothesis h has a certain value *given* the evidence provided. Given that the likelihood is $\Pr(books|female) = 2/3$ and the marginal probability that someone prefers books is $1/2$, we compute the probability that someone is female given the evidence that they prefer books as $(2/3)(3/10)2 = 4/10$. This is an increase in the probability that someone is female from the unconditional probability of being female $(3/10)$. The evidence that someone prefers books to television allowed us to improve our initial estimate that someone is female.

The conditional support for a hypothesis is different than the distribution of a random variable, given whatever knowledge we have about it and given the learning that has taken place. This is referred to as the predictive distribution:

$$\Pr(h) = \int \Pr(h|e)\Pr(e)\,de. \qquad (2.13)$$

This estimate of h represents our knowledge about the value of h considering both the inherent uncertainty about the evidence and the remaining uncertainty about the hypothesis when e is known.

Why Bayes' Rule Matters

One discounts Bayes' rule as "just another mathematical formula" at one's peril. Bayes' rule provides a tool by which experimental evidence can be incorporated into the process of estimating a probabilistic parameter. Consider the situation where we wish to determine whether a document is relevant, given that it has term d. This might be denoted as $\Pr(rel|d)$. While this might be difficult to compute or estimate, one might be able to compute $\Pr(d|rel)\Pr(rel)/\Pr(d)$. The probability $\Pr(d|rel)$ may be computed from a two by two table such as those encountered earlier in the chapter, which contains data from relevance feedback indicating how many documents evaluated by a user are relevant or non-relevant, and how many have the term or don't have the term. When this data is available, along with estimators of other probabilities, the conditional probability, $\Pr(rel|d)$, or a value that is monotonic with it, may be computed. Monotonic functions may be understood informally to go up or down at the same time as the original function; *height* and the *logarithm of one's height* are monotonic; if someone is taller when directly measuring height using inches or meters, one also is taller if one measures the logarithm of the height. This method will be pursued further in Chapter 3; we mention it here to illustrate an application of Bayes' rule to document ranking. By using Bayes rule, we can rank documents by their probability of relevance, given the characteristics of the document.

Extensions to Bayesian Methods

Computing the odds of probabilities provides a great deal of power when making retrieval and filtering decisions. Taking the ratio or odds of two probabilities may often be simpler than computing two separate probabilities, because common factors in the odds may cancel out, simplifying the application of decision theoretic models when ratios or odds are used. The odds for an event x with probability $\Pr(x)$, where $\bar{x} = 1 - \Pr(x)$, are

$$\frac{\Pr(x)}{\Pr(1 - \Pr(x))} = \frac{\Pr(x)}{\Pr(\bar{x})}.$$

Odds are frequently used in decision making when studying the relative chance that a hypothesis is true: $\Pr(h)/\Pr(\bar{h})$. Note that this approach considers the odds that a document is relevant, for example, rather than the straightforward probability that the document is relevant. Computing the conditional odds for a hypothesis makes explicit support both for and against the hypothesis:

$$\frac{\Pr(h|e)}{\Pr(\bar{h}|e)} = \frac{\Pr(e|h)}{\Pr(e|\bar{h})}\frac{\Pr(h)}{\Pr(\bar{h})}.$$

On the right are the prior odds for the hypothesis and on the left are the posterior odds for the hypothesis, given the evidence. The *likelihood ratio* is of particular practical use when working with the odds for and against a hypothesis. Computed as $\Pr(e|h_1)/\Pr(e|h_2)$, the likehood ratio "cancels out" many factors that are irrelevant to the process and focuses on the sufficient statistics that matter.

Sufficient statistics are those factors that characterize a probability distribution. More formally, the likelihood for a random variable can be described fully given all and only the sufficient statistics for that random variable. For example, normally distributed data can be characterized completely by knowing the mean and the variance (spread) of the data. The number of telephone calls initiated in a particular telephone exchange during a small time period is uniquely characterized by the average number of calls initiated during similar time periods. Because the probabilities in the likelihood ratio often have such sufficient statistics, the ratio may turn out to be something as simple as the ratio of the means for the two probabilities.

Taking logarithms of probabilities that are being combined can be useful when interpreting the relationships between the probabilities. For example, assume that there are two independent probabilities and their joint probability, $\Pr(x, y) = \Pr(x)\Pr(y)$. Taking the logarithms of both sides, one can arrive at $\log \Pr(x, y) = \log \Pr(x) + \log \Pr(y)$. This can be used to provide information about the relative contributions of the x and y variables to the joint probability.

The logarithm of the likelihood ratio (often referred to just as the *log likelihood*) can be interpreted as the amount of information available supporting one hypothesis as opposed to another. One can thus derive Bayes' rule in logarithmic form as

$$\log\frac{\Pr(h|e)}{\Pr(\neg h|e)} = \log\frac{\Pr(e|h)}{\Pr(e|\neg h)} + \log\frac{\Pr(h)}{\Pr(\neg h)}.$$

If the logarithms are taken to the base 2, both the likelihood ratio and the prior odds can be compared as to the relative quantity of information each contributes toward the posterior odds, with the results measured in bits.

These equations, relating hypotheses and evidence, contain the likelihood, that portion of the equation where the evidence for the hypothesis enters. The likelihood is a valuable function and appears frequently in decision theoretic and learning models. Sir Ronald Fisher, a leading statistician, noted that "probability is inadequate to express our mental confidence... and that the mathematical quantity which appears to be appropriate for measuring our order of preference among different possible populations [is] likelihood..." (Edwards, 1972).

Because the likelihood and the logarithm of its odds are used so frequently, the notation

$$\lambda = \frac{\Pr(e|h)}{\Pr(e|\neg h)}$$

and similarly

$$\bar{\lambda} = \frac{\Pr(\neg e|h)}{\Pr(\neg e|\neg h)}$$

are frequently used to describe the likelihood that positive or negative evidence will be encountered. One can then define the odds that a hypothesis is true, given the evidence, as

$$O(h|e) = \lambda O(h).$$

It is similarly the case that

$$O(h|\neg e) = \bar{\lambda} O(h).$$

Note that the posterior probability of a hypothesis, given evidence, may be recovered from odds because of the following relationship:

$$\Pr(h|e) = \frac{O(h|e)}{1 + O(h|e)}. \tag{2.14}$$

2.4 ESTIMATING PARAMETERS

Population parameters have values that may be correctly learned if the entire population is examined. For example, the average height for males in the United States at a certain time and date is a fixed value. It is unlikely that anyone knows this value, and the exact value would be very difficult and expensive to measure. As new males are born and others die, the population is constantly changing, as is the average height. Instead of computing the exact value, one may want to make an estimate of this value. Such an estimate is usually much cheaper than determining the population parameter's value and is often sufficiently accurate for decision making purposes. Several different methods exist for estimating these parameters.

LaPlace's Principle of Succession

LaPlace suggests that, if, in n experiments, an observed characteristic is seen x times, then $\Pr(x)$ may be calculated as

$$\Pr(x) = \frac{x + 1}{n + 2}. \tag{2.15}$$

An event understood in the computer age as having a probability of .5, such as flipping a coin, could, using the LaPlace method, result in an estimate of the probability of heads as $\frac{1+1}{1+2} = \frac{2}{3}$ after the flip of a single coin that landed as *heads*. The probability of *tails* after the single flip that resulted in *heads* is $\frac{0+1}{1+2} = \frac{1}{3}$. However, the probability of *heads* plus the probability of *tails* equals 1.

A reasonable estimate of the mean, given some data and no prior information, is

$$\Pr(x) = \frac{x}{n}.$$

When using this estimate, we find that the probability is undefined when $n = 0$. Thus, when no flips have occurred, the LaPlace estimate will still provide a valid probability. The "+1" and "+2" are seen frequently in the retrieval and filtering literature for reasons having to do with producing valid probabilities when $n = 0$.

Method of Moments

The *method of moments* finds the distribution parameters associated with a data set by estimating the population moments (e.g., mean, variance) for the data and then uses these moments in determining the parameters of the distribution. From the data, one can estimate the population mean and variance, the first and second moments about the mean, as well as more advanced moments, such as the third moment (*skew*, or shifting to the right or left) and the fourth moment (*kurtosis*, or "peakedness"). Examining Equation 2.2 shows an exponent of 2 in the numerator of the expression and w indicates that this is the second moment. Using higher valued exponents, one can obtain the skew of a distribution (an exponent and moment of 3) or the kurtosis, with an exponent and moment of 4. By using analytic and numerical techniques, one can find the values for the parameters of a distribution of interest, determining the the parameters that produce distributional moments closest to the moments for the data set.

Maximum Likelihood

The method of *maximum likelihood* is used to provide a measure of central tendency by computing the most frequently occurring value in a data set. In the case of most continuous unimodal distributions, this point has the slope of 0 and may be computed by taking the derivative of the distribution and solving for this derivative being 0. One can then determine the parameters of the distribution that allow the distribution to approximate the true distribution in the dataset.

2.5 BAYESIAN METHODS

Bayesians usually treat knowledge about parameter values as being describable by a distribution representing the degree of belief that a parameter has a particular value. Alternatively, this distribution may represent the relative benefit associated with various beliefs about a parameter. Prior knowledge is then combined with any available evidence, using the likelihood functions, to produce the posterior information about the distribution of possible parameter values. When the posterior distribution and the prior distribution describing the probability of particular parameter values in the likelihood distribution are both of the same mathematical family, differing only by parameter values, the distributions are said to be *conjugate* distributions.

The appropriate use of conjugate distributions is a topic of discussion among those who model real-world phenomena. The use of conjugate prior distributions simplifies modeling greatly. If the prior and posterior distributions are from the same family, then repeated learning can take place with the same type of distribution being produced every time. These systems may offer simplified analytic descriptions of these distributions and relatively simple models of the learning process. When non-conjugate prior distributions are assumed, each learning experience will produce a new type of distribution. Empirical distributions are usually more accurate than conjugate priors, but usually produce unwieldy descriptions of greater complexity and require system-intensive numerical analysis. From the standpoint of modeling, conjugate descriptions are usually preferred if the conjugate priors aren't far from the actual distribution of the dataset.

In practice, it has been found that some conjugate distributions are very flexible and will fit most situations if there is a large enough training set. The application of conjugate distributions to relatively small data sets is problematic. Other, non-conjugate empirical distributions that are not in the family of simple parametric distributions can be used to describe the distribution of occurrences in small datasets, yet these empirical distributions carry their own assumptions and may not be appropriate in some cases.

Estimating the characteristics of a set of documents from a small sample is problematic. One can overlearn the characteristics of the few documents in the sample and fail to generalize about characteristics of the population. Consider a connect-the-dots puzzle, where the underlying shape is a circle. If there are hundreds of dots, connecting the dots with either straight lines or with slight curves will produce something that appears like a circle. If there are only three dots available, whether these dots are connected with straight lines or curves will have a marked impact on the final shape. If there were three dots and we didn't know the underlying shape, one might be able to argue for either the use of straight lines or for curves; this is essentially the kind of problem that is faced when estimating distributions based on a few documents.

Beta Distribution

The most common distribution used to model individual's knowledge about binary parameters, such as those found in filtering and retrieval, is the beta distribution. Lee (1989) notes that "any reasonably smooth unimodal distribution on $[0, 1]$ is likely to be reasonably well approximated by some beta distribution..." The probability of the population parameter having a value p, given that it is beta distributed with parameters a and b, is

$$\beta(p; a, b) = \frac{\Gamma(a + b)}{\Gamma(b)\Gamma(a)} p^{a-1}(1 - p)^{b-1}, \tag{2.16}$$

where the gamma function $\Gamma(x) = (x - 1)!$. The mean of this distribution, the average value for p, is $E(p) = a/(a + b)$.

An alternate parameterization of the beta distribution is sometimes used. The probability of p becomes

$$\beta(p; n, x) = \frac{\Gamma(n)}{\Gamma(x)\Gamma(n - x)} p^{x-1}(1 - p)^{n-x-1}. \tag{2.17}$$

Using this distribution, the mean value for p is $E(p; n, x) = x/n$, and the variance is $V(\beta(p; n, x)) = x(n - x)/n^2(n + 1)$.

The use of the beta distribution is often appropriate when there are large samples. While the distribution will closely capture the characteristics of almost any large sample, it is not obvious how useful it is to assume the beta distribution of p when the sample is small (Weiler, 1965). We will accept it in some cases because of its analytic tractability and the ease with which we can approximately describe the feature occurrences and their prior distribution.

Binary distribution with a Beta prior

The beta distribution can be used as the prior and posterior distribution for a binary distributed variable, using either parameterization of the beta distribution (Equations 2.16 and 2.17). The binary distribution $\mathcal{B}(x|p)$ has the beta distribution as its conjugate prior. The posterior distribution differs from the prior distribution only by its parameter values. Beginning with the prior distribution parameterized as $\beta(p|n, x)$ we find that the posterior distribution may be computed as

$$\beta_f(p|n_f, x_f) = \frac{\beta_i(p|n_i, x_i)\mathcal{B}(x_s|p)}{\int_0^1 \beta_i(p|n_i, x_i)\mathcal{B}(x_s|p)\,dp}$$

$$= \frac{p^{x_f-1}(1 - p)^{n_f-x_f-1}}{\int_0^1 p^{(x_i+x_s)-1}(1 - p)^{(n_i+n_s)-(x_i+x_s)-1}\,dp}.$$

In the case of a single binary item, $n_s = 1$. Here and for the following discussion of conjugate distributions, we use the subscript i to denote the initial values,

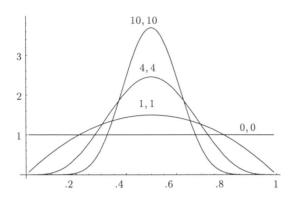

Figure 2.2.   The beta distribution as x_s and n_s increase. Initial knowledge is represented by $x_i = 1$ and $n_i = 1$. The curves peak toward the mean of .5 with number pairs showing x_i and $n_i - x_i$ values.

f to denote the final (posterior) values, and s to denote sample data when clarification is needed. The posterior distribution describing p is the same form of distribution as the prior distribution, where we add x_i to the value x_s, yielding x_f, and we add n_s to the value n_i to yield n_f. Thus, the posterior distribution after incorporating one additional item with value x_s is $\beta_i(p|n_i + 1, x_i + x_s)$. The mean for the prior beta distribution is x_i/n_i and, given relevance information allowing the system to learn, the mean for the posterior distribution $\beta_f()$ becomes $(x_i + x_s)/(n_i + 1)$. Figure 2.2 shows how the probability density near the mean, the actual value for p, increases as x_s and n_s increase.

Gamma Distribution

The gamma distribution (sometimes referred to as the gamma-1 distribution) is often used to describe the distribution of the average for the Poisson distribution, for which the gamma distribution is the conjugate prior. The probability of obtaining this λ is

$$\Gamma(\lambda|t, r) = \frac{e^{-\lambda t}(\lambda t)^{r-1}t}{\Gamma(r)}. \tag{2.18}$$

Given the prior knowledge about λ, denoted as $\Gamma_i(\lambda|t_i, r_i)$, the posterior distribution is $\Gamma_f(\lambda|t_i + t_s, r_i + r_s)$.

The expected value for our knowledge of λ is

$$E(\lambda) = r/t$$

while the variance is

$$V(\lambda) = r/t^2.$$

The parameter r is sometimes referred to as the *shape* parameter and $1/t$ as the *scale* parameter.

Normal Distribution and its Prior

The conjugate prior for the normal distribution's mean is itself a normal distribution. The variance for the sample distribution is assumed to be known or satisfactorily estimated. Assume that the prior distribution describing the mean has mean μ_i and that the sample has n_s data items with mean μ_s. The variance of the process of interest σ^2 is assumed to be known. The posterior distribution is normally distributed with mean μ_f:

$$\mu_f = \frac{n_s \mu_s + n_i \mu_i}{n_s + n_i}.$$

The variance of the posterior distribution describing the mean is

$$\frac{\sigma^2}{n_f},$$

where it is assumed that $n_f = n_s + n_i$.

The conjugate distribution of the variance for the normal distribution is the inverse gamma distribution, or, put differently, the inverse of the variance for the normal distribution has the gamma distribution as the conjugate distribution. This parameterization of the inverse gamma distribution is sometimes treated instead as the inverse chi-square distribution.

2.6 OTHER QUANTITATIVE REASONING SYSTEMS

There are several non-probabilistic quantitative reasoning systems that are widely used in artificial intelligence and are frequently discussed in the research literature. Several of the more popular methods are discussed here.

Fuzzy Logic and Possibility Theory

Probability theory and its related reasoning systems don't seem to represent how people reason or how people remember facts. People are notoriously poor at estimating probabilities and don't reason as suggested by probability when they are given correct probabilities. Fuzzy set theory has been suggested as an alternative to probability theory. Knowledge, it is suggested, may be represented adequately using fuzzy sets, and reasoning may take place using fuzzy operators.

Fuzzy logic is based upon two foundations: fuzzy membership of entities in a set, and a set of fuzzy Boolean operators. Unlike crisp sets to which an entity is, or is not, a member, fuzzy sets allow degrees of membership, representing the degree to which the entity belongs to the set.

Fuzzy operations are similar to Boolean operators. Like Boolean logic, fuzzy logic attempts to model the operations of **and**, **or**, and **not**. The Boolean **and** may be modeled in fuzzy logic as the minimum of the operands. The purpose of **and** in Boolean logic is to limit. If we search for x **and** y, the retrieved documents are limited to those with both x and y. The fuzzy **and** limits by taking the minimum of the operands. The Boolean **or**, on the other hand, expands the retrieved set. Retrieving documents with x **or** y is broader than retrieving documents with the query x. The fuzzy logic operator corresponding to the Boolean **or** is a maximum function. The value for the result is the greatest of the two operands. The fuzzy **not** function may be implemented as one minus the possibility of the event being negated.

Example.   Consider a document database that one wishes to search for books about both birds and their food. Books about birds may be treated as members of the fuzzy set of books about birds to varying degrees, with a book on robins being a member to degree .95, while a book on ostriches is a member to degree .37 and a book on penguins is a member only to degree .2. If each book is also about food for birds to degree .5, then the book on robins and birdfood is a member of the set of desirable books to degree $min(.95, .5) = .5$, where the minimum functions acts as the **and** conjoining birds and birdfood. Similarly, the book about ostriches is a member of the set of books we wish to retrieve to degree $min(.37, .5) = .37$. Obviously, the book on robins has the food as its dominant component, given this query, while the book on ostriches has the ostrich-ness as its dominant component in this environment.

Expected performance may be computed with fuzzy logic when the possibility of statements may be interpreted as the probability of a statement. Given these base probabilities, one can compute the compute the probabilities of combined statements by using maximum and minimum operators.

Dempster-Shafer Reasoning

There is a problem with representing ignorance in probability theory. A uniform distribution can represent either strong belief in equal probabilities or complete ignorance. If $p + \neg p = 1$, then evidence about p provides info about $\neg p$.

Dempster-Shafer reasoning is founded in sets of events, with the probability of a particular set or profile being $m(e)$. Unlike strict probability theory, e doesn't have to be a single state, e.g., e may be a joint state. For example, the probability a document might be about dogs is $m(dogs) = .3$, $m(dogs, cats) = .5$, and $m(rabbits) = .2$. For other possible subjects x, $m(x) = 0$.

A belief function measures the degree of belief in A,

$$Bel(A) = \sum_{B \subset A} m(B).$$

A belief function has the following characteristics:

- The sum of all beliefs in a universe is 1.

- Belief in the null set, nothing, has the value 0.

- The sum of all beliefs in A and $\neg A$ is less than or equal to 1. $\neg A$ is defined as everything but A (Ng & Abramson, 1990).

The *plausibility* of an item is measured as 1 minus the degree of belief that the item is false. A statement that is believed with degree .8 must be disbelieved with degree of less than or equal to .2. Its plausibility must therefore equal or exceed .8.

The belief function equals the probability function m for single items. Thus, $Bel(rabbits) = m(rabbits) = .2$.

For compound items, e.g. *dogs* and *cats*, the belief function equals the sum of the probabilities of the individual items in all their possible groupings. Thus, $Bel(dogs, cats) = m(dogs, cats) + m(dogs) + m(cats) = .8$. This belief in compound items is always greater than or equal to the probability of the compound items.

Because the belief in a statement and the belief that the statement is false must be less than or equal to 1, one can conclude that

$$Pl(S) - Bel(S) \geq 0.$$

One's interest in a degree of truth for a statement has, as its upper bounds, $Pl(S)$ and as its lower bounds, $Bel(S)$.

Two belief values, b_1 and b_2, can be combined easily. First, the joint probabilities of items between both belief functions are computed. If $m(dogs, cats) = .4$ and $m(dogs) = 0$ for one query and $m(dogs) = .2$ for a second query, the combined probability of $m(dogs) = .08$. The sum of all probabilities for a particular item is computed and divided by a normalizing constant, resulting in the combined degree of belief. The normalizing constant K is computed as $K = 1/(1 - C)$, where C represents the degree of null mutual belief, belief that is impossible because the two individual beliefs are unrelated, making the meaning of the product null. The probabilities that document 1 is about *airplanes* and that document 2 is about *gardening* would be multiplied and then added to the C component because the two items don't support each other and their combination provides no support for belief in either item.

Dempster-Shafer theory has some disadvantages when compared to probability theory. For example, Dempster-Shafer doesn't have an inherent inference mechanism, similar to what is provided by Bayes' rule for probability theory, although there are systems based on modifications of the basic concept. Also, falsehood is easily computed with probability theory if a variable is treated as binary and the probability of truth is known. In Dempster-Shafer systems, however, one can't necessarily deduce the degree of falsehood given the degree of truth that is known.

Example.   Consider the relationship between the following variables in a set of documents: Multiple Sclerosis (MS), Lyme Disease (LD), and Other (O).

	MS (.2)	MS, LD (.4)	LD (.3)	Other (.1)
MS (.3)	MS (.06)	MS (.12)	∅ (.09)	MS (.03)
Other (.7)	MS (.14)	MS, LD (.28)	LD (.21)	Other (.07)

The following probabilities are obtained when K is computed as $1/(1-.09) = 1.099$:

MS	0.385
MS, LD	0.308
LD	0.231
Other	0.077
	1.00

Certainty Factors

Originally developed by Shortliffe for use in the MYCIN expert medical system, certainty factors may be easier to use in many circumstances than more formally justified methods and yet have some characteristics desirable in reasoning systems. Certainty factors represent the strength of support that exists for a belief or hypothesis given evidence. Ranging in value from -1 to $+1$, certainty factors provide a mechanism by which the support provided by new evidence may simply be added to prior support for a hypothesis.

To use certainty factors, two initial factors are computed. I, representing the degree to which support for a hypothesis h would increase given evidence e, is computed as

$$I = \frac{max\left(\Pr(h|e), \Pr(h)\right) - \Pr(h)}{max\left(1, 0\right) - \Pr(h)},$$

with values ranging from 0 when the evidence provides no support for the hypothesis to 1 when the evidence completely supports the hypothesis (Shortliffe & Buchanan, 1990; Ng & Abramson, 1990). D, representing the degree to which support for a hypothesis h would decrease given evidence e, is computed as

$$D = \frac{min\left(\Pr(h|e), \Pr(h)\right) - \Pr(h)}{min\left(1, 0\right) - \Pr(h)}.$$

Thus D approaches 1 when $\Pr(h|e)$ approaches 0, while as $\Pr(h|e)$ increases toward $\Pr(h)$, D approaches 0.

The certainty factor CF may now be computed as

$$CF = \frac{I - D}{1 - min\left(I, D\right)}.$$

In a case where evidence increases the likelihood of the hypothesis, $I > 0$ and $D = 0$, then $CF > 0$. When evidence decreases the likehood of the hypothesis and $D > 0$ and $I = 0$, then $CF < 0$.

Van Melle suggests the following function for combining certainty factors (Ng & Abramson, 1990) which avoids problems inherent in earlier attempts (Shortliffe & Buchanan, 1990) where one piece of evidence in a direction could greatly decrease the value of several pieces of evidence in the other direction. Given two CF's, CF_1 and CF_2, the following are used to provide a combined $CF_c(CF_1, CF_2)$:

$$
\begin{aligned}
CF_c &= CF_1 + CF_2(1 - CF_1) && \text{if both } CF_1, CF_2 > 0 \\
&= -CF_c(-CF_1, -CF_2) && \text{if both } CF_1, CF_2 < 0 \\
&= \frac{CF_1 + CF_2}{1 - min\,(\text{abs}(CF_1), \text{abs}(CF_2))} && \text{if one of } CF_1, CF_2 < 0.
\end{aligned}
$$

where abs(x) is the absolute value of x.

The use of methods such as these can result in systems that operate differently than probabilistic systems, often overcoming perceived problems in using probabilistic systems. Fuzzy logic, for example, can combine numeric and logical information. Dempster-Shafer reasoning is more sophisticated, allowing for different forms of evidence to be incorporated into reasoning systems. On the other hand, certainty factors provide probabilistically based reasoning that is easier to implement and has an intuitive appeal, resulting in its use by several successful expert systems.

2.7 SUMMARY

Probabilistic processes produce measurable events, and knowledge about the events can be used to work backwards to learn more about the producing processes. Descriptions of term and document occurrences in different types of documents may be used by models of the text filtering process to produce the best results that may be obtained, when the assumptions of the models are correct. These probabilistic rules and relationships will be used in the next chapter, where basic retrieval and filtering models are considered.

3 SIMILARITY AND RETRIEVAL DECISIONS

It is more important to have beauty in one's equations than to have them fit experiment

—Paul Dirac, Nobel Laureate

Are an elephant and a unicorn similar? Both have four legs and they both have a structure protruding from their faces. There are obvious differences between the two: elephants are "real," have two horns (tusks), a protruding trunk, and are usually thought to be heavier than a unicorn. If they are similar, to what degree or in what manner are they similar? Being able to compute the degree of similarity assumes the existence of a scale ranging from being completely different (whatever that is) to being as similar as possible. We treat the latter as the degree of similarity that an object has with itself. A more qualitative approach to similarity can indicate that there are characteristics that both the elephant and the unicorn possess. Each characteristic, such as having a horn or having four legs, is different, and can be examined and appreciated by itself.

To determine the similarity between text fragments, whether they are documents or queries, requires the texts be represented by features whose degree of similarity can be measured. In most cases, these representations are straightforward indications of term occurrences. Usually, term occurrences are represented by the frequency with which the term occurs, or by a binary indication of whether the term is present in the text fragment or not. Documents may also be represented by the intellectual products of an analysis of the subject or the meaning of the text. For example, when terms such as *house* or *home* occur, they might be represented by the term *shelter,* bringing together different terms referring to the same concept.

The set of representations associated with a text fragment is often referred to as a document *vector* or *profile.* A vector, d, represents a series of n discrete document characteristics with their associated frequencies, $d = \{d_1, d_2, \ldots, d_n\}$. Each vector may be treated as an arrow moving out from the origin, the point with no conceptual subjectness, to a specific point in a subject space, representing the degree to which the text fragment is about each dimension in the space.

While documents are usually represented by a document vector, information needs are often expressed in a specific language designed for queries. These information needs may be expressed as natural language statements of the need or as structured statements, meeting certain syntactic requirements and expressing the information need in a manner consistent with the requirements of a particular filtering model. For example, Boolean queries consist of terms connected with operators, including **and**, **or**, and **not**. The *atoms* of the query, the most basic components, are virtually always of the same type of representation as the atoms within the document representations.

The types of document features may be categorized using a typology commonly found in social science research. Data representing names of things, including answers to a question, such as *yes* or *no*, different colors, such as *red, white,* and *blue,* or *present* or *absent* are referred to as nominal variables. Ordinal data is nominal data that is intrinsically ordered. Values like *good, better,* or *best* have an ordering to them, as would the natural hierarchy for a state university: *state resident, out of state but country resident,* and *foreign student.* Interval and ratio data are real valued variables, where the intervals between variables have a definite meaning, and ratio data has a zero point so that ratios of values are meaningful. An example of interval data is the Fahrenheit temperature scale, while a temperature scale with an absolute zero such as the Kelvin scale, where the lowest possible temperature is the zero point, would be a ratio scale. Retrieval systems need to address the different kinds of features present in data.

Filtering and retrieval systems commonly use only the features in the query when evaluating documents. One often doesn't have knowledge about the role that other terms might play, such as whether they are more or less likely to occur in relevant documents than in non-relevant documents. If a term isn't in the query it is usually assumed that the term will occur at about the same rate in both relevant and non-relevant documents. Query expansion techniques are being developed that increasingly allow terms to be added to a query to increase the system's discrimination ability given the query (Efthimiadis, 1996).

Once representations are available for both a query and the documents to be considered for possible retrieval, the filtering and retrieval systems can determine the similarity between the query and each document. This may take the form of computing the similarity or dissimilarity between the document and the query in a conceptual space that supports the notion of distance.

Document Ordering

Documents are commonly arranged for presentation to the users in a particular order, with those documents expected to be best presented first, and those expected to be of least assistance presented last. The ordering of documents is based upon each document's *retrieval status value* (RSV), computed using similarity or dissimilarity measures, such as distance, or measures that address differences in term frequencies in relevant and non-relevant documents. These measures are computed over the set of document *types*, the possible document profiles. Computing similarity between two document profiles does not depend on the existence of a particular set of text fragments; instead, it represents the relationship between the document profiles (types) rather than between documents themselves (tokens.)

We denote the ordering of x before y as $x \succ y$. Similarly, x being after or at the same ranking as y is denoted as $x \preceq y$. Documents may be weakly ordered or strongly ordered. Strong ordering requires that documents be ordered by a value, and that each document have a value that is greater than the next document in the order or less than the next document. Weak ordering relaxes the requirements that each document be greater than the next (or less than the next) and allows equality of value to exist. When strong ordering is used, equal valued documents (vis à vis the ordering) are not acceptable.

While a decision making system that addresses user preferences about documents will obviously have advantages over one that doesn't, preferences change and are treated irrationally by humans. Our models of document ordering are thus prescriptive, addressing what should occur, rather than descriptive; we make few claims about human behavior or its rationality. For example, Tversky, Slovic, and Sattah (1988) found consistent differences in the results of experiments where users were asked to choose which of two options was best and in other experiments where users were asked to decide what would make two options equivalent. They found that the "more prominent dimension looms larger in choice than in matching." Similar results have been obtained in other research, and other phenomena, such as preference reversals, have been found to occur under certain conditions. Clearly, modeling human preferences will lead to behavior that is sub-optimal or inconsistent in some cases—the retrieval models below will often address how retrieval and filtering should take place, not how it *does* take place when performed by humans.

3.1   CHARACTERISTICS OF SIMILARITY AND DISTANCE MEASURES

A subject bearing object may be understood as representing a point in a space whose dimensions are determined by the characteristics of the object. The conceptual space in which characteristics are represented has n dimensions for a set of n attributes, with one dimension for each attribute (Sneath & Sokal, 1973). The distance between these points may be interpreted as the dissimilarity between the objects. The similarity between these objects may be computed as how "close" the objects are, a factor treated here as in some sense inversely related to distance. One popular relationship is to assume that similarity between two objects is one minus the distance between the two objects, where the maximum distance that can be computed has been normalized to 1.

An examination of spaces capable of containing and representing objects and their characteristic's values begins with the set of real numbers $\mathbb{R}$. If each such number is thought of as a unique point on a line, $\mathbb{R}$ may be thought of as a one dimensional space. $\mathbb{R}^2$ is a two dimensional space, and, to generalize, $\mathbb{R}^n$ is an n-dimensional space. A space may be fully defined by both the dimension of the space and the *metric* or measure used between points in the space.

A *Euclidean space* is the set of all tuples $x_1, x_2, \ldots, x_n$ and $y_1, y_2, \ldots, y_n$ such that

$$\delta(x, y) = \sqrt{\sum_{i=1}^{n}(x_i - y_i)^2}.$$

Williams and Dale (1965) note that there are three ways in which Euclidean properties may be lost. The first is where n in the above formula is not constant. The second is when the n-tuples are constrained, such as being limited to being on the surface of a sphere. The third is when the distance function holds within sets but often fails between sets.

Metrics

A metric is a measure in a particular space, a function having any of a number of traits. The following constraints are all the constraints necessary for a space to be a metric space with a distance metric function $\delta(x, y)$ between two points x and y in the space $\mathbb{R}^n$:

$\delta(x, y) = 0$ for all x and y where $x = y$	(non-degeneracy)
$\delta(x, y) \geq 0$ for all x and y	(non-negativity)
$\delta(x, y) = \delta(y, x)$ for all x and y	(symmetry)
$\delta(x, y) \leq \delta(x, z) + \delta(z, y)$ for all x, y, and z	(triangular inequality)

These constraints may be most easily understood when the metric is the familiar Euclidean distance measure between two points x and y, $\delta(x, y)$. The principle of non-degeneracy claims that there is zero distance between any point and itself and only between a point and itself. This has the effect of preventing two points from being at the same location, where the same location is understood as being those points at distance 0 from a location.

The principle of non-negativity conversely says that if two points are at two locations, which may or may not be different locations, they are at some nonnegative distance. Obviously, if the two locations are different then the distance will be positive, but this is beyond the claim of this axiom.

The principle of symmetry simply states that the distance between a first point and a second is the same as the distance between the second and the first. This requirement for metric spaces need not be the case for all spaces. Consider a space where, instead of distance, the metric represented time or a measure of energy or speed. The movement downwards from one point to another is often different in regards to these measures than the movement upwards between the same two points.

A general rule governing the lengths of sides of triangles, the triangular inequality, assumes that the shortest distance between two points is a straight line (Shreider, 1974).

Varieties of Metrics

If the requirements of a metric are relaxed or changed, other useful categories of similarity and dissimilarity functions may be obtained. In some situations, a function is not a metric, that is, it does not meet the requirements established for metrics. These functions may still serve useful purposes, or may fall into another useful category, meeting requirements for something other than a metric.

If the requirement of non-negativity is dropped, so that it is allowed that $\delta(x, y) < 0$, and all the other requirements of a metric are met, the measure is a *pseudometric*. Dropping the requirement of symmetry from a metric produces a *non-symmetric metric*.

Similarly, if all the requirements of a metric are met but the triangular inequality is dropped, the distance measure is a *semi-metric*. In cases where there are three objects, a, b, and c, and there is a relationship between a and b and a relationship between b and c but there is no known data concerning a relationship between a and c, we may wish to consider a semi-metric. In this situation it is difficult to state that the triangular inequality is or is not met.

When the triangular inequality is dropped as a requirement and replaced with the stronger requirement:

$$\delta(x, y) \leq max\left(\delta(x, z), \delta(z, y)\right) \qquad \text{for all } x, \ y, \ \text{and } z, \qquad (3.1)$$

we have an *ultrametric* (Schweizer & Sklar, 1983).

Set Theoretic Similarity

The above notions of a metric (and its variants) rest on geometric foundations. Other possible structures, such as sets, are useful as foundations for distance and similarity measures. One of the foundations of metric definitions of either distance or similarity is symmetry:

$$\delta(x, y) = \delta(y, x) \text{ for all } x \subset X, y \subset Y.$$

Tversky (1977) has compiled several sets of experimental data showing that humans do not judge similarity as though it were symmetric in some circumstances.

An example from each of two categories of asymmetric relationships may suffice for purposes here. Consider whether you would say that a son resembles his father. Such statements are much more common than statements asserting that the father resembles his son. In general, statements of the form "S resembles L" are seen as asymmetric in many instances when S is a smaller derivative of L.

An experiment was conducted in which 69 subjects were asked whether they preferred the statements "country a is similar to country b" or "country b is similar to country a. Tversky reports that 66 of the subjects preferred the phrase "North Korea is similar to Red China" while only 3 subjects preferred "Red China is similar to North Korea." This great disparity may be taken as evidence that human judgements of "is similar to" are asymmetric. Non-metric axioms for similarity have been developed by Tversky that address similarity measures for qualitative characteristics and set theoretic domains.

3.2 DISTANCE

A variety of distance measures that are consistent with metrics described above may be applied to functions on the points in different spaces. Table 3.1 provides a set of distance measures. We usually assume that data is binary. A distance $\delta(x, y)$ can be understood as a metric or measure that assigns a non-negative number to a set of 2 points x and y (Millman & Parker, 1981). Distances can be computed over many spaces or planes. For example, the Euclidean distance measures the distance between any two points on a flat plane. Similarly, distances may be measured on harder-to-visualize spaces, such as a hyperbolic plane.

The conventional distance in a Euclidean (traditional) space is computed as

$$\delta(x, y) = \left(\sum_i (x_i - y_i)^2 \right)^{1/2}.$$

This is D_1 in Table 3.1. For example, if y represents the origin of a graph, with coordinates $y_1 = 0$ and $y_2 = 0$, and $x_1 = 3$ and $x_2 = 4$, the distance between x and y becomes $\sqrt{3^2 + 4^2} = 5$.

The Minkowski metric (D_2), is computed as

$$\delta(x, y) = \left(\sum_i |x_i - y_i|^r \right)^{1/r}. \tag{3.2}$$

Table 3.1.   Distance measures. All are metrics for positive values.

Label	$\delta(x, y)$	Common names
D_1	$\left(\sum_{i=1}^{n}(x_i - y_i)^2\right)^{1/2}$	Euclidean distance.
D_2	$\left(\sum_{i=1}^{n}|x_i - y_i|^r\right)^{1/r}$	Minkowski metric.
D_3	$max|y_i - x_i|$	Supremum or Dominance metric, as $r \to \infty$ in D_2.
D_4	$\sum_{i=1}^{n}|x_i - y_i|$	City Block metric (Hamming distance) as $r \to 1$ in D_2.
D_5	$\left(\sum_{i=1}^{n}\frac{(x_i - y_i)^2}{(x_i + y_i)^2}\right)^{1/2}$	
D_6	$\sum_{i=1}^{n}\frac{|x_i - y_i|}{(x_i + y_i)}$	Canberra metric.
D_7	$\sum_{i=1}^{n}\frac{|x_i - y_i|}{|x_i| + |y_i|}$	
D_8	$\frac{\sum_{i=1}^{n}|x_i - y_i|}{\sum_{i=1}^{n}|x_i + y_i|}$	
D_9	$\frac{\sum_{i=1}^{n}|x_i - y_i|}{n}$ $\sum_{i=1}^{n} max(x_i, y_i)$	
D_{10}	$\sum_{i=1}^{n}\left(1 - \frac{min(x_i, y_i)}{max(x_i, y_i)}\right)$	

When $r = 2$ the Euclidean Distance (D_1) is obtained. When $r = 1$ for the Minkowski metric (D_2), the *city-block* metric (D_4) is obtained. It is also known as a Manhattan or taxi-cab distance, as well as the Hamming distance.

As $r \to 1$ in the Minkowski metric, the metric approaches an addition function, while as $r \to \infty$ in the Minkowski metric, the formula approaches the dominance metric, in which the function $\delta(x, y)$ returns D_3, the maximum of the two values, x or y. This metric suggests that the distance between X and Y is determined by the distance between the most dissimilar characteristic.

EXERCISE 3.1. [*Easy*] Assume that a query is at location $(3, 4)$, document 1 is at location $(4, 5)$ and document 2 is at location $(3, 3)$. The first number in parentheses represents the spatial location of the query or document in regards to the term *text* and the second number the same for the term *filtering*. What is the Euclidean distance from the query to each document? Which document is "closer" to the query?

EXERCISE 3.2. [*Moderate*] Consider a measure of distance that is the square of the Euclidean distance. What kind of metric (if any) is this?

3.3 SIMILARITY MEASURES FOR NOMINAL DATA

Two objects are *similar* if and only if

1. Each object has a set of observable characteristics,

2. The values for these characteristics are known,

3. A set of characteristics has the same values for both objects, and

4. The above set of characteristics contains either:

   (a) one or more characteristics that are arbitrarily deemed necessary for the two objects to be "similar," or

   (b) there are n or more characteristics with the same values, where n is an arbitrarily selected threshold for "similarity."

The characteristics chosen for identifying similarity are critical to the production of a meaningful measure of similarity. If only one characteristic is to be used for determining similarity and the characteristic *has a trunk* is selected to be used in computing similarity, elephants and unicorns won't be similar, given most reasonable thresholds or sets of characteristics that must be same-valued for similarity to exist. On the other hand, if the characteristic *discussed in children's literature* was to be the sole characteristic considered, both creatures would be considered similar.

The similarity between two objects with measurable characteristics is the "closeness" of the relationship between the two objects. While similarity **S** may be measured as a degree of closeness, it may be unambiguously measured under many circumstances as the inverse of the distance δ between the objects, or $\delta = 1 - \mathbf{S}$.

A similarity measure may take any of a number of different forms, depending on considerations determined by the environment in which the similarity is to be computed. A set of similarity measures is provided in Tables 3.2 and 3.3. A good selection criteria is whether a measure is based on a distance measure that is a metric. However, not all similarity measures are based on metric-based distance measures. These measures may be modified or normalized. For example, the distance between objects may be normalized in many cases by the potential range for each characteristic used in calculating the distance between objects.

One of the most common situations that arises in applications using similarity measures is the need to compute the similarity between two binary variables or between two entities with several distinct binary variables. The notation that is often used is as follows:

Table 3.2.   Similarity measures using d. Measures S_1, S_3, and S_4 are monotonic.

$$S_1 = \frac{a+d}{T}$$

$$S_2 = \frac{a}{T}$$

$$S_3 = \frac{a+d}{T+b+c}$$

$$S_4 = \frac{a+d-b-c}{T}$$

$$S_5 = \frac{a+d}{b+c} \qquad\qquad S_{5a} = \log S_5$$

$$S_6 = \frac{1}{2}\left(\frac{a}{a+b}\frac{a}{a+c}\right)$$

$$S_7 = \frac{1}{4}\left(\frac{a}{a+b} + \frac{a}{a+c} + \frac{d}{d+b} + \frac{d}{d+c}\right)$$

$$S_8 = \frac{ad}{bc}$$

$$S_9 = \frac{\sqrt{ad}-\sqrt{bc}}{\sqrt{ad}+\sqrt{bc}}$$

$$S_{10} = \frac{ad-bc}{ad+bc}$$

$$S_{11} = \frac{4bc}{T^2}$$

$$S_{12} = \frac{ad}{\sqrt{(a+b)(a+c)(d+b)(d+c)}}$$

$$S_{13} = \frac{ad-bc}{\sqrt{(a+b)(a+c)(d+b)(d+c)}}$$

$$S_{14} = \frac{a+\sqrt{ad}}{a+b+c+\sqrt{ad}}$$

$$S_{15} = \frac{2(a+d)}{T+a+d}$$

		Vector X present	Vector X absent	
Vector	present	a	b	$a+b$
Y	absent	c	d	$c+d$
		$a+c$	$b+d$	$T.$

Here the similarity between entities are characterized by the the number of features that are present in both, present in one but not the other, or present in neither. Consider the case where we have a universe of 26 features $A, B, \ldots, Z$, a query with features F, G, and H, and a document with features F and G. The table will have the following values:

		Query Features		
		present	absent	
Document	present	$a = 2$	$b = 0$	$a + b = 2$
Features	absent	$c = 1$	$d = 23$	$c + d = 24$
		$a + c = 3$	$b + d = 23$	$T = 26.$

Once the values have been developed for such a table, they may be used in computing similarity measures such as those in Table 3.2 ($\mathbf{S}_1 - \mathbf{S}_{15}$) and Table 3.3 ($\mathbf{S}_{16} - \mathbf{S}_{22}$).

Simple Match

A simple method for computing similarity is provided by the *simple match* which counts the number of similarly valued variables. This may be normalized to provide the percentage of variables that have equal values:

$$\mathbf{S}_1 = \frac{a+d}{a+b+c+d} = \frac{a+d}{T},$$

with the lowest value equaling 0 and the highest value equaling 1. Consider matching two vectors of binary variables as:

	Data Values						Value
Data X	1	0	0	0	0	1	
Data Y	0	0	0	1	0	1	
Effect	↓	↑	↑	↓	↑	↑	0.667

Here we show the direction of the effect of individual variables on the final value, shown on the right. Given the data in this form or in tabular form,

		Vector X		
		1	0	
Vector	1	$a = 1$	$b = 1$	2
Y	0	$c = 1$	$d = 3$	4
		2	4	$T = 6$

the value of the simple match function may be computed. A variable with the same value in both entities can be seen to increase the simple match value, while variables with differing values will lower the value of the simple match function.

The expected value for the simple match is:

$$E(\mathbf{S}_1) = \frac{(a+b)(a+c) + (d+b)(d+c)}{T^2}.$$

The simple match coefficient can be easily related to the notion of Euclidean distance by noting that, for binary data, $D_1 = \sqrt{1 - \mathbf{S}_1}$. This assumes that the nominal variables are assigned values 1 and 0. The value $1 - \mathbf{S}_1$ can be understood as the proportion of attributes with unequal values. These values contribute to the distance between the two vectors X and Y, unlike those attributes that are equal in X and Y and do not increase the distance between X and Y.

Variations on the Simple Match

In some cases one may wish to emphasize matches for terms that are present, with less emphasis on matching absences of terms:

$$\mathbf{S}_2 = \frac{a}{T}.$$

The d factor is still included in the denominator, but does not affect the numerator for $\mathbf{S}_2$.

Its operation may be viewed as

	Data Values						Value
Data X	1	0	0	0	0	1	
Data Y	0	0	0	1	0	1	
Effect	↓	↓	↓	↓	↓	↑	0.167

The expected value of this is (Sneath & Sokal, 1973):

$$E(\mathbf{S}_2) = \frac{(a+b)(a+c)}{T^2}.$$

The similarity may also be computed emphasizing dissimilarities as (Sneath & Sokal, 1973):

$$\mathbf{S}_3 = \frac{a+d}{a+d+2(b+c)}.$$

This value is monotonic with the value produced by the simple match.

Its operation may be viewed as

	Data Values						Value
Data X	1	0	0	0	0	1	
Data Y	0	0	0	1	0	1	
Effect	↓	↑	↑	↓	↑	↑	0.500

The expected value of this measure of similarity is (Sneath & Sokal, 1973):

$$E(\mathbf{S}_3) = \frac{(a+b)(a+c) + (d+b)(d+c)}{b+c+T^2}.$$

Similarity may be emphasized as:

$$\mathbf{S}_{15} = \frac{2(a+d)}{2(a+d)+b+c} = \frac{2(a+d)}{T+a+d}.$$

It functions as:

	Data Values						Value
Data X	1	0	0	0	0	1	
Data Y	0	0	0	1	0	1	
Effect	↓	↑	↑	↓	↑	↑	0.800

The expected value of this unnamed coefficient is (Sneath & Sokal, 1973):

$$E(\mathbf{S}_{15}) = \frac{2\left((a+b)(a+c) + (d+b)(d+c)\right)}{T\left((a+b) + (a+c)\right) + 2(d+b)(d+c)}.$$

Unlike the simple match, which provides the ratio $(a+d)/T$, the odds form eliminates the (a+d) from the denominator, resulting in

$$\mathbf{S}_5 = \frac{a+d}{b+c}$$

with the lowest value equaling 0 and the highest value equaling ∞. This is not the α cross product found in statistics, which is ad/bc. The odds form operates as:

	Data Values						Value
Data X	1	0	0	0	0	1	
Data Y	0	0	0	1	0	1	
Effect	↓	↑	↑	↓	↑	↑	2.000

If a similar measure is desired with the values instead ranging from $-\infty$ to ∞, one might use the logarithm of the former measure:

$$\mathbf{S}_{5a} = \log \frac{a+d}{b+c}.$$

3.4   SIMILARITY EXCLUDING JOINT ABSENCES

Table 3.3 provides a set of similarity measures that ignore the value of d. This has the effect of not considering joint absences of a characteristic when measuring similarity. The author and his wife are thus neither more nor less similar because neither is a stegasaurus or a unicorn. Emphasis instead is placed on those characteristics that occur in at least one of the entities.

Jacquard's similarity measure is the simple match function with the d factor removed. It is the proportion of characteristics occurring in either entity that occur in both,

$$\mathbf{S}_{16} = \frac{a}{a+b+c},$$

with the lowest value approaching 0 and the highest value equaling 1.

It operates as:

Table 3.3. Similarity measure not using d, joint absences. Measures S_{16}, S_{17}, S_{18}, and S_{19} are monotonic.

$$S_{16} = \frac{a}{a + b + c}$$

$$S_{17} = \frac{a}{b + c} \qquad\qquad S_{17a} = \log S_{17}$$

$$S_{18} = \frac{a}{a + (b + c)/2}$$

$$S_{19} = \frac{a}{a + 2(b + c)}$$

$$S_{20} = \frac{a}{\sqrt{(a + b)(a + c)}}$$

$$S_{21} = \frac{a}{max(a + b, a + c)}$$

$$S_{22} = \frac{a}{min(a + b, a + c)}$$

	Data Values						*Value*
Data X	1	0	0	0	0	1	
Data Y	0	0	0	1	0	1	
Effect	↓			↓		↑	0.333

Given two vectors X and Y, the set theoretic definition of the Jacquard measure is

$$S_{16} = \frac{\|X \cap Y\|}{\|X \cup Y\|}.$$

If there are m items in the intersection of X and Y, $\|X \cap Y\|$, and n items in the union, $\|X \cup Y\|$, then the Jacquard coefficient can be computed as m/n. This may also be understood as a ratio of probabilities. Limiting ourselves to a universe Z of z items, the probability of being in the intersection becomes m/z while the probability of being in the union, $X \cup Y$, is n/z. This simplifies to the earlier m/n, suggesting that the ratio of probabilities can be seen as equivalent to the Jacquard measure.

The expected value of the Jacquard measures is (Sneath & Sokal, 1973):

$$E(S_{16}) = \frac{(a + b)(a + c)}{T((a + b) + (a + c)) - (a + b)(a + c)}.$$

When the dissimilarity between entities is to be emphasized while otherwise accepting the premises of the Jacquard similarity measure, one may use the following:

$$\mathbf{S}_{19} = \frac{a}{a + 2(b + c)}$$

with the lowest value approaching 0 and the highest value equaling 1.
This measure operates as

	Data Values						Value
Data X	1	0	0	0	0	1	
Data Y	0	0	0	1	0	1	
Effect	↓			↓		↑	0.200

The expected value of this similarity coefficient is (Sneath & Sokal, 1973):

$$E(\mathbf{S}_{19}) = \frac{1}{(2T/(a + b) + (2T/(a + c) - 3)}.$$

Dice's coefficient

Dice's coefficient is computed as

$$\mathbf{S}_{18} = \frac{2a}{2a + b + c},$$

with the lowest value approaching 0 and the highest value equaling 1. The
operation is:

	Data Values						Value
Data X	1	0	0	0	0	1	
Data Y	0	0	0	1	0	1	
Effect	↓			↓		↑	0.500

The value of Dice's coefficient has been shown to be monotonic with the value
of the Jacquard similarity measure (Sneath & Sokal, 1973) and the ranking as
the same as that found with the Jacquard measure (Losee, 1990). It also may
be understood in set theoretic terms as

$$\mathbf{S}_{18} = 2\frac{||X \cap Y||}{||X|| + ||Y||},$$

with the lowest value approaching 0 and the highest value equaling 2 where the
notation $||X||$ represents the size of the set X (Van Rijsbergen, 1979).
The expected value of Dice's coefficient is

$$E(\mathbf{S}_{18}) = \frac{2}{\frac{T}{a + b} + \frac{T}{a + c}}. \tag{3.3}$$

The odds form of the Dice coefficient is

$$S_{17} = \frac{a}{b + c},$$

with the lowest value approaching 0 and the highest value approaching ∞. It operates as:

	Data Values						Value
Data X	1	0	0	0	0	1	
Data Y	0	0	0	1	0	1	
Effect	$\downarrow$			$\downarrow$		$\uparrow$	0.500

If a measure is desired with similar characteristics but with a range approaching $-\infty$ to ∞, one might take the logarithm of the former measure thus:

$$S_{17a} = \log \frac{a}{b + c}.$$

EXERCISE 3.3. [*Easy*] Consider a query containing terms x and y. Assume document a has terms x and z and document b has terms x and y. Which similarity measures will produce a lower similarity value between the query and document a than between the query and document b? Which measures would give an equal similarity value?

3.5 BOOLEAN RETRIEVAL

Boolean queries express information needs as a logical statement specifying the characteristics needed by a document for it to be retrieved. If the query has the value *true* for a document, the document is retrieved or passed through the filter. Each term in the query has the value *true* only if the term occurs in the document being examined. A Boolean logic is used to combine the truth values for individual terms using the operators **and**, **or**, and **not**.

Traditional Logics

A proposition or statement in propositional logic is either an assertion that a simple fact is true, such as "the car is red," or a combination of simple facts and logical operators. A fact must be in the form of a statement; the use of only a noun, e.g., *car*, or an adjective, e.g., *red*, is insufficient. In arithmetic, valid statements might look like "3", "2+1", or "(3+2+1)/2". In arithmetic, simple facts are numbers, while the operators are the symbols representing operations like addition, subtraction, multiplication, and division.

In logic, five operators are commonly used. They are "$\neg$", representing logical negation of the following statement, and four binary operators which operate on the propositions before and after them (like arithmetic addition). The four binary operators are "$\wedge$", representing logical *and* and conjunction, "$\vee$", representing logical *inclusive or* and disjunction, "$\supset$", representing logical implication, and "$\equiv$", representing the statement "if and only if." The latter is also referred to as a biconditional operator; if $p \equiv q$, then $p \supset q$ and $q \supset p$.

Operators and statements may be combined into an infinite number of logical expressions.[1]

It needs to be emphasized that these operators take propositions as arguments. As Sainsbury (1991) points out, the statement

Tom and Mary got married

is not equivalent to

Tom got married and Mary got married.

The latter is more adequately represented by propositional logic, while the conjunction in the first is far more difficult to represent directly in propositional logic.

The following truth values exist for the above logical operators, given the varying truth values for p and q presented on the left:

p	q	$\neg p$	$p \wedge q$	$p \vee q$	$p \equiv q$	$p \supset q$
T	T	F	T	T	T	T
T	F	F	F	T	F	F
F	T	T	F	T	F	T
F	F	T	F	F	T	T

Not all these operators are necessary for a logic to exist. One can propose a logic with axioms defining logical negation and disjunction (or), and can then proceed to derive the rules for conjunction. This is done through the definition

$$(p \wedge q) \equiv \neg(\neg p \vee \neg q).$$

Bertrand Russell chose negation and disjunction as the basis for the logic developed in *Principia Mathematica*. Similarly, conjunction and negation can be used to produce disjunction:

$$(p \vee q) \equiv \neg(\neg p \wedge \neg q).$$

Other operators, such as implication, can be defined in terms of these lower operators:

$$p \supset q \equiv \neg(p \wedge \neg q) \equiv \neg p \vee q.$$

Boolean Logic

Boolean logic, first fully developed by George Boole, provides additional rules for manipulating logical expressions. *Commutivity*, $(p \wedge q) \equiv (q \wedge p)$ and $(p \vee q) \equiv (q \vee p)$ suggests that the order in which conjunctions and disjunctions are

---

[1] The disjunction of several expressions may be treated as the logical sum of the truth values, where 1 represents *true* and 0 represents *false*, and where the logical sum has no "carrying," and thus never exceeds the value 1 (*true*). Likewise, the conjunction of several expressions is the logical product of these expressions, thus, a 0 (*false*) for any of the values produces a 0 (*false*) for the result, as in multiplication.

written is unimportant. The law of *associativity* implies that the grouping by
parentheses, the order by which operations take place, is as follows:

$$p \wedge (q \wedge r) \equiv (p \wedge q) \wedge r$$

and

$$p \wedge (q \wedge r) \equiv (p \wedge q) \wedge r.$$

The laws of *negation* state that $(p \wedge \neg p)$ is tautologically or necessarily false
and $(p \vee \neg p)$ is tautologically true. The following laws can be deduced from
these and the earlier laws:

$$\neg(\neg p) \equiv p \qquad \text{(Complementation)}$$
$$p \wedge p \equiv p \text{ and } p \vee p \equiv p \quad \text{(Idempotency)}$$
$$p \wedge (p \vee q) \equiv p. \qquad \text{(Absorption)}$$

Boolean Operators and Retrieval

The Boolean expression x will retrieve documents with term x present. We
may think of the presence of term x as being *true*.

The Boolean **and** operator takes the truth value of both its arguments and
returns the value *true* for the expression if, and only if, the two operators each
have the value *true*. Thus, when using an expression such as

text **and** filtering

in a filter, only documents containing both the terms *text* and *filtering* are
passed through the filter. Documents with only one, or neither of the terms,
but not both, will not be passed. The **and** is used to limit the set of retrieved
documents, adding additional requirements that must be met if a document
is to be retrieved. This is different than an informal natural language notion
of "and" which suggests that by **and**ing terms to the query, one is adding
additional ways a document could be retrieved, that it is broadening the search,
when it is always the case that **and**ing terms narrows the search.

The Boolean connective **or** is used when listing alternatives, any of which
provide grounds for the expression to be labeled *true* and which would justify
its retrieval. For example, house pets might be listed as

dog **or** cat **or** bird **or** fish,

and a document containing any one of these terms would be retrieved. Unlike
and, **or** is used to expand the set of retrieved documents, which goes contrary
to some informal interpretations of the natural language term "or."

When a particular term or expression should not occur in retrieved documents, the **not** is used. If we wished to exclude documents about cold, one
would phrase a query as

not cold.

In practice, negation is less commonly used than the **and** and **or** connectives.

Queries in Normal Form

When comparing Boolean queries, it is often necessary to place the queries into a common form. One can always express a Boolean expression as a disjunction of conjunctions, that is, a set of conjunctions, each containing a set of atoms or their negation that are conjoined, with these conjoined expressions being disjoined. Such an expression is in *disjunctive normal form* (DNF). For example, one might have an expression such as

hot **or** (temperature **and** (**not** cold))

which is in DNF, with the **or** as the highest level Boolean operator in the expression.

Another commonly used form for Boolean expressions is Conjunctive Normal Form (CNF), a conjunction of disjunctions. All expressions may be converted into a series of disjunction of atoms or their negations which are then conjoined. An expression in DNF such as

(a **and** b **and** c) **or** (d **and** (**not** e))

may be converted to CNF as:

(a **or** d) **and** (a **or** (**not** e)) **and** (b **or** d) **and**

(b **or** (**not** e)) **and** (c **or** d) **and** (c **or** (**not** e))

Expressions in CNF may be viewed as conjoining *hyperfeatures*, with each disjunction supplying a list of the possible characteristics of the hyperfeature (Bookstein, 1985). The CNF expression may then be easily treated probabilistically if the hyperfeatures are treated as statistically independent of each other (Losee & Bookstein, 1988). Relevance feedback may be incorproated into this Boolean system by treating it as probabilistic relevance feedback regarding the frequencies of hyperfeatures in relevant and non-relevant documents. The CNF expression may also be viewed as a fuzzy expression and the minimum function used to combine the frequencies of the hyperterms in the document.

EXERCISE 3.4. [*Easy*] What would be a satisfactory Boolean query that would retrieve this book but would not retrieve books addressing analog filters or books about information theory?

3.6   ANGULAR DISTANCE AND VECTOR RETRIEVAL

The use of the angle between two vectors as indicative of the degree of similarity or dissimilarity may have been first suggested formally by Bhattacharyya in 1946. He suggested that the cosine measure itself can serve as a coefficient of association.

The angle between two vectors $\vec{X}$ and $\vec{Y}$, with the lengths of vector X denoted as $|X|$ and similar notation for the length of Y, may be computed in a Euclidean

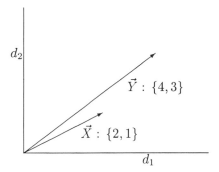

Figure 3.1.   Distance between two vectors as the angle between the vectors.

space as

$$\cos\theta = \frac{\displaystyle\sum_{i=1}^{n} x_i y_i}{|X| \cdot |Y|}, \tag{3.4}$$

where the numerator represents the *inner product* of X and Y. The denominator ($|X| \cdot |Y|$), a *normalization factor*, normalizes the cosine for the length of the vectors, where $|X|$ is the length of X, with similar notation for Y.

For the vectors in Figure 3.1, the inner product of X and Y may be calculated as $4 \cdot 2 + 3 \cdot 1 = 11$ and, with a normalizing factor of $5\sqrt{5}$, the cosine of .984 is obtained.

Similarly, the cosine of Y and Y can be computed by first calculating the inner product, 5, which is then divided by the normalizing factor $\sqrt{5} \cdot \sqrt{5}$, yielding a cosine similarity value of 1.

When the binary values 1 and 0 are used for the attribute values, the cosine measure can be understood by examining its relationship to measures using the contingency table above. The cosine measure may be rewritten as

$$\frac{a}{\sqrt{a+b} \cdot \sqrt{a+c}}, \tag{3.5}$$

where a represents the number of items in the intersection $\|X \cap Y\|$ and $\sqrt{a+b}$ and $\sqrt{a+c}$ represent the lengths of the binary vectors or sets X and Y. Thus, if the vectors being compared are binary, containing only values 0 and 1, and where 1 may be interpreted as set membership and 0 as non-membership, the cosine similarity between them may be computed as

$$\frac{\|X \cap Y\|}{\sqrt{\|X\|}\sqrt{\|Y\|}}, \tag{3.6}$$

where $\|Z\|$ represents the number of items in set or vector Z.

The value of the cosine for a 6 dimensional vector is

	Data Values						Value
Data X	1	0	0	0	0	1	
Data Y	0	0	0	1	0	1	
Effect	↓	↓	↓	↓	↓	↑	0.500

EXERCISE 3.5. [*Easy*] What is the cosine of the vector representing the title "Information Retrieval and Filtering" and the vector representing the title "Information Theory and Practice"?

Mixing Vector and Boolean Models

One model of Boolean and fuzzy retrieval that relates them to the vector model is the *p-norm* model (Greiff, Croft, & Turtle, 1997; Schweizer & Sklar, 1963; Salton, 1984; Fox & Winett, 1990) in which the weight for a set of n **and**ed terms is computed as

$$1 - \left(\frac{1}{n} \sum_{i=1}^{n} (1 - w_i)^p \right)^{1/p} ,$$

where p, $1 \leq p \leq \infty$, is a constant for the model and where term i has weight w_i. Similarly, the weight for a set of n **or**ed terms is computed as

$$\left(\frac{1}{n} \sum_{i=1}^{n} w_i^p \right)^{1/p} .$$

When p approaches 1, the results are similar to those obtained with the vector model, while when p approaches ∞, the results are similar to those obtained with traditional Boolean operators.

3.7   PROBABILISTIC RETRIEVAL

The decision to retrieve a document, consistent with statistical decision theory, is modeled by the probabilistic retrieval model. It assumes that the expected cost associated with retrieving a document should be minimized for the searcher (Maron & Kuhns, 1960; Van Rijsbergen, 1979; Robertson, Van Rijsbergen, & Porter, 1981; Bookstein, 1983; Salton & McGill, 1983). The decision rule is as follows:

> Retrieve a document if the expected cost of retrieving the document is less than the expected cost of not retrieving the document.

This may be expressed as: retrieve a document only when

$$EC_{retr} < EC_{\overline{retr}},$$

the expected cost of retrieving a document is less than the expected cost of not retrieving a document. Frequently a bar will be placed over a variable to indicate the negation of the variable or the variable being applied to non-relevant documents. The expected cost of retrieving a document is estimated probabilistically as

$$EC_{retr} = \Pr(rel|d)C_{retr,rel} + \Pr(\overline{rel}|d)C_{retr,\overline{rel}},$$

where $\Pr(rel|d)$ denotes the probability that a document is relevant given that it has a set of features with frequencies d and the cost of retrieving a relevant document is denoted as $C_{retr,rel}$, and $C_{retr,\overline{rel}}$ represents the cost associated with retrieving a non-relevant document. A similar expression is used for the expected cost of retrieving a non-relevant document, with $\overline{retr}$ representing not retrieved.

Note that each expected cost represents the average of two costs. Each expected cost is the expected cost of an action, with the two costs being averaged between the two possible states-of-nature associated with the action. Thus the expected cost of retrieving a document is the weighted average of the cost of retrieving (the action) a relevant (the state-of-nature) document and the cost of retrieving (the action) a non-relevant (the state-of-nature) document.

The retrieval rule becomes: retrieve a document with characteristics d if and only if

$$\Pr(rel|d)C_{retr,rel} + \Pr(\overline{rel}|d)C_{retr,\overline{rel}}$$
$$< \Pr(rel|d)C_{\overline{retr},rel} + \Pr(\overline{rel}|d)C_{\overline{retr},\overline{rel}}.$$

This may be transformed to produce a rule suggesting that a document be retrieved if and only if

$$\frac{\Pr(rel|d)}{\Pr(\overline{rel}|d)} > \frac{C_{retr,\overline{rel}} - C_{\overline{retr},\overline{rel}}}{C_{\overline{retr},rel} - C_{retr,rel}} = constant. \tag{3.8}$$

That is, a document with characteristics d should be retrieved if the odds that it is relevant exceeds a particular constant.

These costs are related to the kind of results that are desired by the searcher. While all costs have an impact on the cost constant that must be exceeded if a document is to be retrieved, some costs are more important in certain types of situations. The user who wishes to retrieve almost all documents on a topic (a *high recall* search) assigns a high loss to not retrieving a relevant document. Those who are very concerned about having a set of documents which are almost all relevant to the needs of the end user (a *high precision* search) will assign a higher cost to retrieving a non-relevant document than will searchers more willing to accept more non-relevant documents in the set of retrieved documents.

The cost associated with retrieving a relevant document may be the dominant cost when making the retrieval or filtering decision. It represents the worth of a good document to the end user. On the other hand, the cost of not retrieving a non-relevant document is usually very small, with the cost of not looking at something that is irrelevant to their needs being insignificant.

When using Equation 3.8, one can decide whether to retrieve a document or not depending on whether the ratio on the left hand side of the equation exceeds the cost constant on the right hand side. In many retrieval situations, the user may find it difficult to provide the costs necessary to compute the cost constant, and the system *ranks* the documents by the value of the left hand side of the equation. The documents are then presented to the user in this ranked order. The user then decides to *stop* retrieving documents when it seems best for them to do so (Kraft & Lee, 1979; Kraft & Waller, 1981). Ranking allows us to use this formula instead of determining the value for the costs.

The primary difference between filtering documents, as the documents move across a network, and retrieving documents from a database is that the user retrieving documents may receive a ranked list of documents, since the system has the opportunity to examine all the documents that will be presented to the user before giving them to the user. A filtering system, on the other hand, must make the decision to present the documents to the user based solely on the query or user profile as it then exists and the characteristics of the document. There is no possibility of storing documents which are then ordered so that the best document is first, the second best is second, and so forth. This is one reason that filters that do not batch documents together may be seen as inferior to filters that can batch documents together, with the documents being ranked within a batch.

Assuming Feature Independence

Using the ratio derived in Equation 3.8 above, we find that

$$\frac{\Pr(rel|d)}{\Pr(\overline{rel}|d)} = \frac{\Pr(d|rel)}{\Pr(d|\overline{rel})} \frac{\Pr(rel)}{\Pr(\overline{rel})}. \tag{3.9}$$

If we assume conditional term independence, that is, term frequencies in relevant and in non-relevant documents are statistically independent, we may then compute

$$\prod_{i=1}^{n} \frac{\Pr(D_i = d_i|rel)}{\Pr(D_i = d_i|\overline{rel})} \frac{\Pr(rel)}{\Pr(\overline{rel})}. \tag{3.10}$$

The $\prod$ symbol is used to represent the product of a set of values, i.e.,

$$\prod_{i=1}^{3} x_i = x_1 x_2 x_3.$$

We can drop the $\Pr(rel)/\Pr(\overline{rel})$ component in Equation 3.10, which is constant for any given query and independent of the characteristics of a particular document, to produce

$$\prod_{i=1}^{n} \frac{\Pr(d_i|rel)}{\Pr(d_i|\overline{rel})}.$$ (3.11)

This formula can be used to rank documents with independent terms by first computing the probability of having a particular feature value, given the appropriate relevance class, and then ranking the documents by this value.

In addition to dropping parts of formulae that don't affect the ranking of documents, one sees scholars dividing the above expression (and formulae like it) by the same expression but with $d = 0$, providing a degree of normalization (Bookstein, 1983; Robertson & Walker, 1994).

Binary Independence Assumptions

We saw earlier that if terms are binary, we may estimate the probability of a term occurrence d as

$$\Pr(d|rel, p) = p^d (1 - p)^{(1-d)},$$

where p is the probability of having a term in a relevant document. We will similarly treat the same distribution but for non-relevant documents with probability q that a non-relevant document has a term. We then find that Equation 3.11 becomes

$$\prod_{i=1}^{n} \frac{\Pr(d_i|rel)}{\Pr(d_i|\overline{rel})} = \prod_{i=1}^{n} \left(\frac{p_i/(1-p_i)}{q_i/(1-q_i)} \right)^{d_i}.$$

Taking the logarithm of the right hand side of this expression, we arrive at the document ranking formula:

$$RSV_d = \sum_{i=1}^{n} d_i \, \log \frac{p_i/(1-p_i)}{q_i/(1-q_i)},$$ (3.12)

the retrieval status value (RSV) or document weight. This may be viewed as sum of the weights for each term. Each term's weight is computed as the discrimination value of the term or the degree to which the term discriminates between relevant and non-relevant documents. This value is multiplied by the frequency of term i, d_i, for each term, and these values are summed, producing the document weight.

Example.   Assume that we have six documents, each with two features, and the relevance values given in Table 3.4. We find that for the two terms, we have parameters $p_1 = 2/4$, $p_2 = 3/4$, $q_1 = 1/3$, and $q_2 = 1/3$. Natural logarithms

Table 3.4.   A set of sample documents for analysis.

Doc.	d_1	d_2		Doc	d_1	d_2	
A	1	0	relevant	B	1	1	relevant
C	0	1	relevant	D	0	0	non-relevant
E	0	1	relevant	F	1	0	non-relevant
G	0	1	non-relevant				

are used here, and the discrimination values for the terms are, respectively,

$$\ln \frac{(1/2)/(1-(1/2))}{((1/3))/(1-(1/3))} = \ln 2 = .693$$

and

$$\ln \frac{(3/4)/(1-(3/4))}{(1/3)/(1-(1/3))} = \ln 6 = 1.792.$$

Given these values, the ranking (with tied documents grouped in parentheses) is given from right to left, along with the weights, as

$$(B\ 2.485)\quad (C\ E\ G\ 1.792)\quad (A\ F\ .693)\quad (D\ 0).$$

Because document B has both features, its weight is computed as 1 times the discrimination value for the first term (.693) plus the frequency of the second term, 1, multiplied by the discrimination value for the second term, 1.792, yielding 2.485. Similarly, since document D has neither term, its weight becomes 0.

EXERCISE 3.6. [*Easy*] Assume that document B above is non-relevant. What will be the discrimination values for the two features and what will be the ordering of the documents?

Two Poisson Independence Assumptions

Using Equation 3.12, term frequencies can be incorporated into a retrieval model if we consider them to be Poisson distributed (Harter, 1975a, 1975b; Bookstein, 1983; Raghavan, Shi, & Yu, 1983; Losee et al., 1986; Srinivasan, 1990; Margulis, 1993). This assumption is not met exactly by term frequencies in natural language text, but is close to the actual state of nature, making it a useful assumption to make. Assuming that terms are distributed so that they have probability of having d occurrences, given the average frequency of λ for the appropriate relevance class,

$$P(d|\lambda) = e^{-\lambda}\lambda^d/d!,$$

we may compute

$$RSV_d = d_i \ log \ \frac{\lambda_i}{\bar{\lambda}_i},$$

where λ is the average frequency for relevant documents and $\bar{\lambda}$ is the average frequency for non-relevant documents.

The application of this model is an exercise below.

Normally distributed features

If the values held by a given variable are assumed to be normally distributed, one may wish to use the discrimination value for each term as suggested by Bookstein (1982). The retrieval status value for a normally distributed term is

$$RSV_d = d_i \left[\left(\frac{\mu_r}{\sigma_r^2} - \frac{\mu_n}{\sigma_n^2} \right) - \frac{d_i}{2} \left(\frac{1}{\sigma_r^2} - \frac{1}{\sigma_n^2} \right) \right],$$

where μ_r and μ_n represent the means for relevant and non-relevant documents, respectively, and σ_r^2 and σ_n^2 represent the variances for relevant and non-relevant documents, respectively.

Note that when the variances are the same for both distributions, the ranking becomes the same as the ranking obtained with the cosine of the document vector and the vector of $(\mu_r - \mu_n)/\sigma^2$ values. Bookstein (1982) shows that for a large family of probabilistic distributions (members of the exponential family), the ranking obtained with the probabilistic model is the same as (or similar to) the ranking obtained with the cosine similarity measure.

EXERCISE 3.7. [*Moderate*] Assume that we have the following data set:

Doc.	d_1	d_2		Doc	d_1	d_2	
A	1	0	relevant	B	3	1	relevant
C	0	2	relevant	D	0	0	non-relevant
E	0	1	relevant	F	1	0	non-relevant
G	0	1	non-relevant				

and that terms are assumed to be Poisson distributed. What are the discrimination values for the two features? What will be the ordering of the documents using these weights?

EXERCISE 3.8. [*Moderate*] Assume that for the previous exercise the discrimination values for both terms are computed using the binary discrimination formula, where a term is considered present if it has a frequency greater than or equal to 1. What would be the discrimination value for both of the terms? What would be the ranking if the binary discrimination values are multiplied by the non-binary term frequencies?

EXERCISE 3.9. [*Research Problem*] Consider the development of a negative binomial term independence model. Why might this model be better than the binary independence or two Poisson independence models considered here? Why might it be inferior?

3.8   THE BAYESIAN LEARNING MODEL

The parameters for the binary independence and two Poisson independence retrieval models may be estimated using Bayesian methods. Because the problem of ranking documents probabilistically is based on statistical estimation (Cooper, 1994), using the sophisticated estimation methods provided by Bayesian techniques can lead to a rich model of retrieval. If we have binary independent data, we may treat the prior knowledge about the probability of p or q as though it is described by a beta distribution. Using Equation 2.17 we can progressively estimate the values of p as we gain more information about the characteristics of relevant documents (Bookstein, 1983). Similarly, the prior information about the Poisson parameter λ may be described by the gamma distribution, consistent with Equation 2.18.

Relevance Feedback

While documents may be retrieved based upon the relationships between the documents and the initial queries, *relevance feedback* provided by searchers, indicating which documents they find relevant and which they find non-relevant, can provide additional information about user interests. This may be used by a retrieval system to make increasingly accurate estimates of probabilistic parameters and thus improve performance. Early forms of relevance feedback were batch oriented, taking sets of retrieved relevant documents and sets of retrieved non-relevant documents and developing new weights based on the characteristics of the documents in these sets (Spink & Losee, 1996). Relevance feedback may also be interpreted as producing a new query. Query expansion or modification produces a new query in the query language of the system. This new query may then be submitted to the system to retrieve additional documents (Efthimiadis, 1996). More recent feedback techniques have allowed users to effectively learn from each individual document when it is examined, using formally based models of knowledge (Losee, 1988; Allan, 1996).

Bayesian methods for incorporating relevance feedback do so by representing knowledge about the characteristics of relevant and non-relevant documents as parameters for prior distributions. For notational simplification, we treat here a single term, dropping the subscripts indicating the term in many cases. To use multiple terms, the term weights are summed as above.

We initially estimate $p = x_{Ri}/n_{Ri}$, where x_{Ri} represents the conceptual initial (i) number of documents having the term in question from among n_{Ri} conceptual relevant (R) documents. The values x_{Ri} and n_{Ri} represent our prior beliefs as to the characteristics of relevant documents. A similar set of parameters, x_{Ni} and n_{Ni}, represent our prior knowledge about non-relevant (N) documents.

When relevance feedback is supplied, sample data (subscripted with s), may be used to update p and q. After relevance feedback has been incorporated, we can estimate $p = (x_{Ri} + x_{Rs})/(n_{Ri} + n_{Rs})$ and $q = (x_{Ni} + x_{Ns})/(n_{Ni} + n_{Ns})$.

Using these values, the weight for the binary independence model becomes

$$
d\log\left(\frac{\dfrac{(x_{Ri}+x_{Rs})/(n_{Ri}+n_{Rs})}{1-(x_{Ri}+x_{Rs})/(n_{Ri}+n_{Rs})}}{\dfrac{(x_{Ni}+x_{Ns})/(n_{Ni}+n_{Ns})}{1-(x_{Ni}+x_{Ns})/(n_{Ni}+n_{Ns})}}\right).
$$

As more relevance feedback is supplied by the user, the values for x_{Rs}, n_{Rs}, x_{Ns} and n_{Ns} are updated, modifying the weights, for individual terms, based on the user's expressed preferences. When using this technique, *we are learning which documents the user identifies as interesting* and any pattern meeting the assumptions of the model can be learned. This model of relevance feedback *does not depend on having a satisfactory or philosophically satisfying definition of relevance.* Relevance need only be an arbitrary label assigned to documents. The system learns about documents with these labels.

There are problems remaining to be solved with these feedback models. One is how to learn from a biased sample (typically the "best" documents the system can find) rather than from a random sample of documents of each relevance category, a basic assumption of virtually all statistical models. Another problem is what parts of a retrieved document we should learn from. Clearly, retrieving an encyclopedia with a relevant passage and learning that all the material in the encyclopedia is relevant will mislead a learning procedure (Allan, 1995; Callan, 1994). What part of a document should be used in feedback? Should the feedback be scaled down gradually as one moves away from the "most significant" part of the document? This problem will clearly need to be solved as systems increasingly use differing sources and sizes of documents and the demand for improved relevance feedback increases.

Example.   Consider again the sample data in Table 3.4 that was ranked under the BI model. We will again rank documents by their retrieval status value, the document weight, and, in the case of a *tie* between several documents with the same weight, the document whose name is alphabetically earliest will be retrieved first. Let us assume that $x_{Ri} = x_{Ni} = 1$ and that $n_{Ri} = n_{Ni} = 2$ and thus all p and q values will be estimated as $1/2$ for both features. The documents initially will be assigned weights as follows:

A, F	1×0.0	$+$ 0×0.0	$=$	0.0
B	1×0.0	$+$ 1×0.0	$=$	0.0
C, E, G	0×0.0	$+$ 1×0.0	$=$	0.0
D	0×0.0	$+$ 0×0.0	$=$	0.0

The first document to be retrieved will be A, the alphabetically first document name from those documents with the tied highest weights (of 0). Since A is relevant, we now improve our knowledge about our p parameters, which describe the characteristics of relevant documents. Since feature 1 in A is a 1, we will modify $p_1 = 1/2$ to $p_1 = (1+1)/(1+2) = 2/3$ by adding 1 (the feature value) to x_{Rs1} and 1 (because one document was retrieved) to n_{Rs1}. Here the third subscript represents the term number. Because the second feature has value 0, we add the value 0 to x_{Rs2} and add 1 (for one document) to n_{Rs2}, making $p_2 = (0 + 1)/(1 + 2) = 1/3$.

When the remaining unretrieved documents are reranked using our new knowledge (new parameter values), one obtains:

F	1×0.693	+	0×0.0	=	0.693
B	1×0.693	+	1×0.0	=	0.693
C, E, G	0×0.693	+	1×0.0	=	0.0
D	0×0.693	+	0×0.0	=	0.0

The alphabetically earliest name for a document with the highest weight (0.693) is B. When relevant document B is retrieved, with its 1s for both features, we update parameters x_{Rs1}, n_{Rs1}, x_{Rs2}, and n_{Rs2} by adding a 1 to each of them. These update our estimates for parameters p_1 and p_2 to 3/4 and 2/4, respectively, making the discrimination values for the two terms 1.79 and 0, respectively. The unretrieved documents may be reranked, producing a ranking that is more reflective of the optimal ranking, as the estimates of p_1 and p_2 continue to approach the true probabilities and approach the true average frequencies for the binary term frequency in relevant and non-relevant documents.

EXERCISE 3.10. [*Easy*] Assume that $x_{Ri} = 1$, $n_{Ri} = 2$, $x_{Ni} = 1$, and $n_{Ni} = 2$. For a single term query, assume that two relevant documents are retrieved with the term, one relevant document without the term, and one non-relevant document with the term. What are the current estimates of p and q for this term after the relevance feedback has been incorporated into the parameter estimates? Given these estimates, what would the weight be for a document with a term frequency of 1? For a document with a term frequency of 0?

Two Poisson Independence and Feedback

When raw term frequencies are available beyond the binary indicators of term presence or term absence, using the two Poisson retrieval model may improve performance. We begin by assuming that terms are Poisson distributed, and the gamma distribution is assumed to represent our prior knowledge about the average term frequency λ and $\bar{\lambda}$ for each relevance category. We initially estimate $\lambda = x_{Ri}/n_{Ri}$, where x_{Ri} represents the initial number of term occurrences from among n_{Ri} conceptual documents. The values x_{Ri} and n_{Ri} represent our prior beliefs as to the characteristics of relevant documents. A similar set of parameters, x_{Ni} and n_{Ni} represents our prior knowledge about $\bar{\lambda}$, the average term frequency in non-relevant documents.

When relevance feedback is supplied, sample data (subscripted with s) is used to update λ and $\bar{\lambda}$. After relevance feedback, we can estimate $\lambda = (x_{Ri} + x_{Rs})/(n_{Ri} + n_{Rs})$ and $\bar{\lambda} = (x_{Ni} + x_{Ns})/(n_{Ni} + n_{Ns})$. When terms are Poisson distributed, λ and $\bar{\lambda}$ may exceed one when a single document is examined, unlike the case with the binary independence model, where each has its maximum at 1.

Using these values, the term weight for the two Poisson independence model becomes

$$d \log \left(\frac{\dfrac{(x_{Ri} + x_{Rs})}{(n_{Ri} + n_{Rs})}}{\dfrac{(x_{Ni} + x_{Ns})}{(n_{Ni} + n_{Ns})}} \right).$$

This is summed for each term, producing the document weight.

Term Frequency and Document Length

The Poisson model may be modified to incorporate the length of a document. Terms will obviously occur with greater expected frequency in larger documents than in shorter documents, all other things being held constant. The parameter λ in the Poisson model may be replaced with $l_d \lambda$, where l_d is the number of units (length) in document d and λ is the average number of term occurrences per text unit. When there are l_d text units in a document, the probability of having d term occurrences in the document becomes

$$\Pr(d | \lambda, l_d) = \frac{e^{-\lambda l_d} (\lambda l_d)^d}{d!}$$

passage length. Other methods are available that compensate for document length (Singhal, Buckley, & Mitra, 1996).

EXERCISE 3.11. [*Research Problem*] Develop a model of retrieval with relevance feedback (assuming binary term distributions) that takes into account the fact that the first document retrieved is probably the best relevant or non-relevant document, the second the second best, and so forth.

3.9 RELATED WEIGHTS

The *inverse document frequency* (IDF) weighting has a long history as an effective weighting technique in filtering and retrieval systems. Computed as $-\log n/N$, where N is the number of documents and n the number with the term in question, a rare term that occurs in 1 out of 100 documents would have a weight of $-\log .01$, while a common term occurring in 99 of 100 documents would have an IDF weight of $-\log .99$. The highest IDF weights are assigned to the rarest terms, while common terms have weights near 0.

We can understand one reason that the IDF weighting scheme works as well as it does by considering its relationship to the decision theoretic model. The n/N component may be treated as q in the binary independence model. The probability that a non-relevant document has the term in question is q and, if almost all documents are non-relevant, this approximates n/N (Croft & Harper, 1979)

Consider a document weight based on $\log[p/(1-p)/q/(1-q)]$. Let us hold the p constant, such as when the p knowledge is the same for all terms. The decision theoretic weight would then be proportional to $\log q/(1-q)$. This is clearly related to $\log q$ and both will rise and fall as q changes.

The IDF weight may be combined with a term frequency model like the Poisson model by using a TF (term frequency) times IDF term weight. This "TF-IDF" weight, as it is popularly referred to, uses the term frequency, rather than just the binary presence or absence of the term, enabling it to provide an increased document weight when terms occur more frequently and a lower weight when the term occurs only a few times.

The IDF and TF-IDF weights require no relevance information to compute and apply. However, using information supplied by query terms about relevant documents may improve retrieval and filtering performance (Losee, 1988).

EXERCISE 3.12. [*Easy*] Assuming that the information about relevant documents is held constant, would would be an effective term weight like the IDF weight if terms were Poisson distributed?

3.10   UPPER BOUNDS

The *upper bound* or best-case for a search represents the best expected results that can be obtained given the query terms and the limiting of document terms used in ranking to those mentioned in the query. We assume that the system can rank documents only by those features in the query, using the various probabilities associated with those queries. Our estimate of the upperbounds assumes no direct knowledge of the relevance status of individual documents. Upper bounds can thus use full probabilistic knowledge, such as might be learned or estimated in a system, but don't have direct knowledge about the relevance of individual documents. This stops one from selecting all the relevant documents and placing them at the head of a list, but does allow one to place those documents most probably relevant at the beginning of the list of documents.

Note that our notion of the upper bounds is explicitly probabilistic, in that it addresses the best *expected* results. There may be individual results that are better than this; our focus here is on expected results and our definition of upper bounds is defined in terms of expected results.

Theorem 3.1 (Upper bounds ordering) *Consider that there are n distinct document profiles, $g_1, g_2, \ldots, g_n$, representing the order in which groups (or a part of the final group) are retrieved, and there are $N(g_i)$ documents in group g_i and $R(g_i)$ relevant document in the group. For all g_i, g_j, and $i \leq j$, when documents are ordered so that*

$$\frac{R(g_i)}{N(g_i)} \geq \frac{R(g_j)}{N(g_j)},$$

the ordering is the upper bounds for this set of documents.

An ordering consistent with this constraint places any group with a higher density of relevant documents before all others with a lower density. When groups of documents with the higher densities are placed as far forward as possible in the list of documents, the upper bounds are obtained. This places documents into a weak ordering by decreasing expected precision, since the component $R(g_i)/N(g_i)$ is monotonic with expected precision.

As a corollary to this, the lower bound ordering is obtained simply by reversing the ranking of the upper bounds.

Note that in a practical retrieval situation using relevance feedback, when two groups have the same ratio, it may be best to select documents from the group which one has learned the least about if we anticipate retrieving many more documents and we assume that it is best to learn the same amount about each feature rather than to learn the same amount but have this unequally distributed over the terms. This will result in a more even learning about the characteristics of relevant (and non-relevant) documents to be retrieved in the future.

Upper Bounds for Boolean Queries

When using a Boolean query, the ranking is limited to two groups, those documents retrieved and those not retrieved and thus conceptually ranked lower. One can retrieve all the documents from group g_1 through group g_j placed in upper bounds order with an optimal query by writing the query which retrieves all and only the documents in group g_1 as well as all and only the documents in group g_2, and so forth, until all and only the documents in group g_j have been retrieved. If the set of features in g_i are denoted as $\{g_i\}$ and the individual features in this group are **and**ed together, we may retrieve the upper bound (best) set of documents through group g_j by **or**ing together the sets $\{g_1\} \vee \{g_2\} \vee \{g_3\} \vee \cdots \vee \{g_j\}$.

Example.   Consider 4 groups, the first defined by having terms x and y, the second by x and not y, the third by not having either, and the fourth by not having x and having y. The upper bounds query through the third group g_3 is:

$$(x \text{ and } y) \text{ or } (x \text{ and } (\text{ not } y)) \text{ or } ((\text{ not } x) \text{ and } (\text{ not } y)).$$

3.11   BROWSING

Documents may be retrieved by emulating how one browses from one book on a shelf to another, moving from one aspect of a subject to another. Libraries organize documents using classification systems such as the Dewey and Universal Decimal Systems or the Library of Congress Classification system, which ideally place related documents near each other. At the end of groups of documents, there are usually "breaks" as the classification system jumps to a different subject area, but, except for these jumps, the similarities between adjacent documents are useful for browsing. By placing similar materials nearby, clas-

Table 3.5. Reflecting Gray code. Note that the right half is in the reverse decimal order from the left half. The Gray code on the right hand side is the same as that on the left (for the same row) with a 1 placed as the left most digit.

Decimal	Binary	Gray Code	Decimal	Binary	Gray Code
0	0000	0000	15	1111	1000
1	0001	0001	14	1110	1001
2	0010	0011	13	1101	1011
3	0011	0010	12	1100	1010
4	0100	0110	11	1011	1110
5	0101	0111	10	1010	1111
6	0110	0101	9	1001	1101
7	0111	0100	8	1000	1100

sified collections often provide the searcher with useful materials about which the user hadn't thought about seeking (Morse, 1970; Boll, 1985; Baker, 1986; Marchionini, 1987; Cover & Walsh, 1988; Huestis, 1988; Losee, 1993a, 1993b).

The linear ordering provided by a classification system supporting browsing can be generated automatically. Hypermedia objects can be automatically linked and browsed using similar methods (Losee, 1997a). If features in documents are binary, the optimal arrangement for a dense set of documents covering all possible topics is provided by a *reflected binary Gray code*. A Gray code (Table 3.5) is a counting system that enumerates in such a way that adjacent representations in the system differ by exactly one bit (Flores, 1956; Gilbert, 1958; Hamming, 1986; Conway, Sloane, & Wilks, 1989).

Expected Dissimilarity

The difference between a document and another document is the sum of the differences between the features. When using an information theoretic difference, the Expected Informational Dissimilarity (EID) is

$$EID = \sum_{i=1}^{n} |d_{i,1} - d_{i,2}| \left(-p_i \log p_i + (1 - p_i) \log(1 - p_i) \right),$$

where p_i is the probability that feature i occurs in a document, $d_{i,j}$ is the binary value of feature i in document j, and n is the number of document profiles. When documents are ordered so that the EID for a term decreases as one moves from left to right within the Gray code representation of a document, the EID between adjacent documents is decreased for the entire set of documents, providing a superior ranking to that provided by other Gray codes. The expected mutual information measure (EMIM) discussed in Chapter 7 may also be used to study the distance between numbers encoded with the Gray code.

3.12   SUMMARY

Retrieval and filtering models often determine the similarity or the distance between the query or stated information need and the document. Many distance and similarity measures have the desirable and mathematically well-behaved characteristics of a metric, while others are more *ad hoc*. Some models directly address the decision to allow documents to pass through a network filter or to be retrieved. Boolean queries state explicitly what a document must have (or cannot have) if it is to be retrieved or passed through a filter. The probabilistic retrieval models similarly use a query which serves as initial information about what might be in a relevant document. Given probabilistic models, relevance feedback may be used to increase the system's knowledge about what the user wants.

Next, we turn to how one measures the quality of these document retrieval procedures.

4 MEASURING PERFORMANCE

To be in hell is to drift; to be in heaven is to steer.

—George Bernard Shaw

Using precise measures of filtering performance allows a system user, manager, or designer to:-

- describe what happened in the past,

- predict what will happen in the future,

- determine whether one method is better than another, and

- determine whether performance is at its best.

Existing filtering and retrieval systems may be evaluated using any of a number of measures developed over the past several decades. These performance measures may be computed from past activities or they may be used to predict future activities. The ability to accurately predict the operation of a system can lead to increased managerial control, improved use of the system, decreased costs for all involved, and an increased understanding of how the system works. System purchasers, managers, and users can then make reasoned decisions about what circumstances should exist if one is to choose to use a particular retrieval algorithm, database, query language, or search engine.

Text filtering and retrieval system performance is most commonly expressed as one or more precision and recall values, both measures indicating characteristics of the retrieved document set and the retrieval system. These measures, in turn, assume that documents are judged, or can be judged, as relevant or non-relevant. Other measures or characteristics of text filtering performance may be used, each having its own interpretations and strengths and weaknesses.

Many measures of retrieval performance are used to address average system performance, often using the mean or median of a set of individual performance measures to evaluate system performance. A set of composite measures themselves may be averaged, or the individual values that are used by each of the composite measure may be aggregated. In studies addressing what *could* be obtained, one might measure or compute the best results obtained for each of a number of queries, providing information about the upper limits of performance. In other cases, system designers might wish to avoid the worst-case situations, and thus might pay attention to the worst-case performance in each of a number of trials, trying to improve this number, perhaps accepting mediocre performance overall in a risk-avoiding effort to not encounter the worst-case performance.

It must be remembered that averages are not the same as the raw data on which the averages are based. In some unusual circumstances the mean carries most of the information that an experimenter would need about the data set, but in most other situations, information is being discarded when the average is presented to the user in lieu of presenting the raw data or additional parameters of the dataset. Relying on the average as an indicator of performance may mask interesting patterns and characteristics in the results, yet it provides a useful single number characterizing a dataset.

The Average Search Length (ASL), the average position of a relevant document in the ordered list of documents, will be considered in depth here both for its use as a retrospective measure of performance for a single query and because of its analytic predictability and tractability. With this easy to interpret measure, searchers and researchers can predict performance, using methods discussed in later chapters. By predicting performance, we mean stating what is expected from the system, based upon rational principles. Using ASL, the performance of specific retrieval models may be studied, described, and predicted by computing the degree to which a particular ranking model is optimal, combined with a particular ordering of documents. This degree of optimality is dependent on a number of factors, and identifying the factors that cause one retrieval model to be intrinsically better or worse than another can lead to an understanding of the operation and characteristics of superior document manipulation systems.

Retrieval vs. Filtering System Performance

While retrieval and filtering conceptually are very similar, there are some differences in their function and thus in the concerns to be addressed when measuring system effectiveness. Document retrieval systems retrieve one or more documents at a time, with the user making the decision when to stop the retrieval process. Filtering programs are faced with a continuous stream of documents, requiring immediate decisions about whether to forward or discard each individual document arriving at the filter.

Most document retrieval systems have the assistance of a human being to stop the system from providing additional documents when the user considers searching no longer worthwhile. Traditional Boolean retrieval systems exactly describe the characteristics of documents to be presented to the user, but most non-Boolean systems rank documents in order of their decreasing likelihood of use, allowing the user to continue examining documents until the searcher decides to halt the process. Filtering systems continue to present documents to the user as long as the documents arriving at the filter meet the specification established by the user (referred to here as a query). Filter users may revise their query, or may request that the filter cease sending them documents; this form of stopping is somewhat different in implication than the expected intervention by the searcher in a document retrieval system. In practice, filtering systems often have conservative queries so that the user isn't presented with too many documents, while a text retrieval system may present a much larger number of documents to the end user, who may easily stop the document examination process at any point.

4.1   RETROSPECTIVE PERFORMANCE MEASURES

Retrieval system quality may be measured with respect to many factors. Some measures of performance are affected primarily by the matching procedure used, while some measures address other factors, such as the speed of the hardware or the size or quality of the document database. Users prefer to have the system respond as rapidly as possible, minimizing the user's waiting time. System costs are often lower when hardware and software can respond more rapidly.

Users also may evaluate documents for their perceived quality. For example, measures of the variety, timeliness, collection comprehensiveness, and the accuracy of documents in the collection can represent collection quality. System usability also affects performance. Some systems are easier to use than others. Easier systems may have simpler query languages, require less training, or have better interfaces for providing relevance feedback. Simpler interfaces may increase the chances the user will get what they want, increasing system accuracy and increasing system speed as the chance for mistakes is decreased.

Relevance

The function of a filtering or retrieval engine is to present to the user documents worth examining, while not presenting material to the searcher that is estimated to be of little or no use. Documents that have some usefulness to the end user are often referred to as *relevant*. Relevance may be judged in several different ways. Relevant documents may be relevant to everyone, or they may be judged relevant to a specific searcher. Documents may be judged either clearly relevant or clearly non-relevant, or those above a particular cutoff on a scale may be considered relevant. A document may satisfy an information need, in part or in whole, and may be considered relevant if it satisfies this need in some respect. The role of a text filtering or retrieval system is to obtain relevant documents, satisfying an expressed information need (the query or the filter) or anticipating the future needs of a user or the future usefulness of one or more documents.

As was noted in Chapter 1, objective and topical understandings of relevance, emphasizing that relevant documents have an objective or topical relationship with a query, are the most common notions of relevance. Objective relevance usually results in binary relevance judgements, i.e., the document is either relevant or not relevant to the information need expressed by the query. This may be expanded beyond the simple binary notion of relevance to a multivalued set of judgments. For example, a document might be judged

1. necessary for the query presenter,

2. of some use to the query presenter, or

3. of no use to the query presenter.

Other, more elaborate, linear relevance scales have been proposed. Our discussion below will focus on binary relevance.

In filtering and retrieval systems, users often find a document relevant for a variety of reasons beyond topicality. A document that all agree is "about" the topic of the query may be rejected by the user because it has been read already, or because it is in a language unknown to the reader, because it has too many formulae, or is in poor condition, or for any number of other personal reasons.

A full appreciation of relevance is desirable for developers of experimental databases and for those trying to extrapolate from experimental results to the results expected in a commercial environment. However, the analytic model of retrieval performance does not produce a set of results from which commercial results will be extrapolated, as do most experimental simulations. The filtering professional will merely apply the analytic model to the parameters of a commercial database, rather than to the parameters of a small, experimental database. Performance is predicted below based on whatever relevance judgements a searcher happens to apply to a set of documents. These techniques are not dependent on a particular philosophical stance about relevance, requiring only that the judgments are binary, and even this assumption is dropped for a model predicting retrieval performance consistent with continuous relevance.

Relevance feedback that is used by the models in Chapter 3 requires only that the user say that a document is "relevant" – this can be merely a label that can represent any of a wide number of phenomena. The system learns the characteristics of those documents the user labeled *relevant*, whether they are truly relevant, in a philosophical sense, or not. It can then present documents to the user ordered by the probability that they would be *labeled* as *relevant*. Whether they are in fact truly or philosophically relevant is beside the point; the system learns to recognize how the user labels, and *the user's reasons and motivations for labeling are irrelevant* to the system and its performance.

Locating Relevant Documents

Finding the relevant documents and determining their number can be difficult to perform accurately. The number of relevant documents may be determined directly, by counting the number of relevant documents, or indirectly,
by randomly sampling from the database and then extrapolating the percent
of relevant documents in the sample to the number of relevant documents in
the database. This method is prone to inaccuracies for large databases, as the
size of the random sample may need to be very large to contain even a single
relevant document. In this case, the estimator would be very crude and prone
to large errors.

A second approach is to attempt to retrieve all the documents that might
be relevant. This may be done using a number of different techniques and then
combining the results, with the hope that, between all the different methods,
all the relevant documents will be found. For example, documents on *Chapel
Hill, NC* might be located by searching for the terms *Chapel Hill*, or the point
on a map where the town is located might be touched, with the system retrieving documents on neighboring or encompassing areas like *Orange County*, the
surrounding governmental body. Often, not all the relevant documents will be
found even when using a range of techniques, but these techniques will usually
locate all the relevant documents that would be found by most other related retrieval techniques, and thus can serve as the basis for retrieval experiments. We
refer to these documents as the set of *relevant retrievable documents,* and suggest that documents that can't be retrieved using any known retrieval method
should not influence measures of system performance.

Recall

Given a database of documents, with a subset of these being relevant and a
subset of the database being retrieved, a reasonable measure of performance
is the percent of the relevant documents that have been retrieved. This is
related to the overlap between the subsets of retrieved documents and relevant
documents. The *recall* measure is computed as:

$$recall = \frac{||\{retr\} \cap \{rel\}||}{||\{rel\}||},$$

where $\{retr\} \cap \{rel\}$ represents the intersection of the set of retrieved documents
and the set of relevant documents, $||\{retr\} \cap \{rel\}||$ represents the size of the set
of retrieved relevant documents, where $\{rel\}$ is the set of relevant documents
and $\{retr\}$ the set of retrieved documents. If 10 documents were retrieved and
were relevant and there were 40 relevant documents in the database, we might
say that recall was .25, that is, one quarter of the relevant documents had been
retrieved.

Recall may also be interpreted probabilistically as $\Pr(retr|rel)$, the proba-
bility that a document has been retrieved given that it is relevant.

Precision

The quality of the retrieved set of documents may be studied with the *precision* measure, the percentage of documents in the retrieved set that are relevant:

$$precision = \frac{||\{rel\} \cap \{retr\}||}{||\{retr\}||},$$

where $||\{retr\}||$ represents the size of the set of relevant documents retrieved. The value $||\{rel\} \cap \{retr\}||$ is the size of the set of relevant documents retrieved, making precision the number of relevant documents retrieved divided by the total number of documents retrieved. Precision may be interpreted probabilistically as $Pr(rel|retr)$, the probability that a document in the set of retrieved documents is relevant.

When all documents retrieved are relevant, the precision is 1, the highest value that may be obtained for the measure. When no relevant documents are in the set of retrieved documents, precision is 0.

Note that neither precision nor recall directly addresses the non-retrieval of non-relevant documents.

Example.   Consider a set of documents, retrieved (from left to right) as follows:

r r n r r n n r ,

where r is a relevant document and n is a non-relevant document. Note that there are 8 documents and 5 relevant documents. When the second document is retrieved, the recall is 40%, as two of the five relevant documents have been retrieved. The precision at this point is 100% since both (all) documents retrieved are relevant. By the time that 100% recall has been obtained, when all the documents have been retrieved, the precision has dropped to 5/8, or 67.5%.

Other Measures

Fallout, the portion of the non-relevant documents that are retrieved, is computed as

$$fallout = \frac{||\{\overline{rel}\} \cap \{retr\}||}{||\{\overline{rel}\}||}.$$

While the fallout measure is not widely used by practitioners, it serves as a complement to the precision measure. Those using economic models that assign a loss to retrieving non-relevant documents but assign no loss to retrieving relevant documents can find fallout a useful tool for studying performance.

The density of the relevant documents in the database is measured as

$$generality = \frac{||\{rel\}||}{||\{all\}||}, \tag{4.1}$$

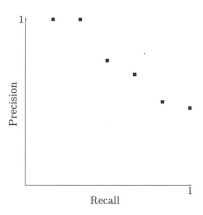

Figure 4.1.   Recall and precision points for retrieving documents.

the number of relevant documents for a query divided by the number of documents in the database, the percent of documents in the database that are relevant to a given query. When the average generality measure for the queries presented to a database decreases, smaller numbers of documents are relevant to each query and the database itself can be said to cover a greater number of topics and thus be less "general" and more specific.

EXERCISE 4.1. [*Easy*] Assume that there is a database with 1000 documents, 100 of which have been retrieved. Half of the retrieved documents are relevant but only ten percent of the documents in the database are relevant. Compute precision, recall, and generality.

EXERCISE 4.2. [*Moderate*] Consider the precision, recall, fallout, and generality measures. Define each one of these in terms of the other three, when possible.

Precision-Recall Graphs

It may be helpful to study the progress of an individual search or of the average progress of a group of searches by studying the precision and recall at a number of different points. This is usually done at each change of recall level, such as when the system moves from 25% recall to 50% recall. An example set of recall and precision values can be derived from a document sequence such as:

r r n r n r n n n r n n n r r n n.

There will be $\|\{rel\}\|$ (7) relevant documents, with x (let us assume 5) unique document profiles producing at most $x + 1$ (i.e., 6) different precision levels.

Precision-recall graphs are probably the most frequent form of presentation for retrieval performance data. Recall is usually plotted on the x axis and precision on the y axis. The data may be plotted as a set of points. Figure 4.1 shows the set of data points representing the retrieval performance for the data above.

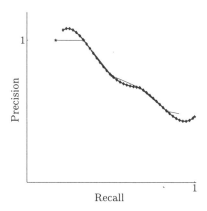

Figure 4.2.   Recall and precision points with smoothing and interpolation.

It is clear that in many cases we want to be able to consider the differences between different aspects of performance, such as low recall vs. high recall. Low recall searches are often found among searches by those who want "some" information on a topic, possibly a few books or articles. In some cases this results in finding everything that has been published on the topic, but most commonly the few documents are just a small fraction of the documents addressing the topic of interest. We may consider most searches by the "general public" to be low recall searches. High recall searches are performed most commonly in academia, patent work, and law, where the searcher needs to find everything that has been written on a topic or that addresses a technique or phenomenon.

It may be desirable to connect the data points representing either individual searches or the points representing average searches. Figure 4.2 shows the data in Figure 4.1 interpolated (the dashed lines). The curve in Figure 4.2 represents the plot of a function that was fit to the data and then sampled over the desired recall range. It also illustrates some of the problems of this "curve fitting" approach, which shows precision at impossible values for the top left part of the figure. For the bottom right portion of the curve, we see that as recall approaches 1 that precision is increasing at the end. This is obviously an artifact of the curve fitting process, with the dashed line representing the connection of precision-recall data points. Commonly available software can produce continuous curves from a set of data points and may be useful producing precision-recall curves.

To estimate the value for a point between two data points for which we have data, it is necessary to interpolate. For example, if there are three relevant documents, precision is only directly available at the 33%, 66%, and 100% recall levels. If precision statistics were desired for the 50% recall level, for example, it would be necessary to estimate a precision value based on the precision values for recall levels of 33% and 66%. Given two recall-precision points a and b with precisions P_a and P_b, and an intermediate precision point

P_x such that $R_a \leq R_x \leq R_b$ from recall levels R_a to R_b, we may compute

$$P_x = P_a + (P_b - P_a)\frac{R_x - R_a}{R_b - R_a}.$$

That portion of the difference in precision between points a and b that lies between point a and point x (computed as the difference between the gap from point x to a, normalized by the width of the gap from b to a) is added to the precision for point a. If we begin with 60% precision at the 33% recall level and 40% precision at 66% recall, we find that if x is the 50% recall point, then $P_x = .50$.

EXERCISE 4.3. [*Easy*] Assume that a set of documents are retrieved in the order:

r n r r n n n r n.

Graph a precision recall curve for this data.

EXERCISE 4.4. [*Easy*] Assume that at the recall level of 25%, precision is .8, and at 50%, precision is .5. What is the interpolated precision at the 30% and 40% recall levels?

4.2 SINGLE NUMBER MEASURES

Single number measures of filtering and retrieval performance are often used to represent degrees of performance. By taking the mean or median of a set of performance values, for example, one can summarize the performance process. Single number measures might also be used to represent the average of the best and the worst performance results obtained.

A number of single number measures have been used or proposed to evaluate filtering and retrieval systems (Kraft & Bookstein, 1978). Single number measures can be compared and the method with the best value can be selected. This avoids the problem that arises when comparing precision-recall curves, where one method may be better at high recall and another method superior at low recall.

Measures may focus on the performance of a retrieval or filtering system for the individual user or for the system as a whole. An average may be computed from each user's search performance, or from specific data aggregated from all system operations used to compute system performance. Most text filtering and retrieval system measures are understood to be measures of performance for the individual user; when performance for the system as a whole is intended, it is usually referred to as a *system oriented* measure (Salton & McGill, 1983).

Most measures address raw performance. One different form of measure, the Asymmetric Uniqueness measure (AU), computes the percent of a set of relevant documents unique to that particular retrieval method (Tenopir, 1985):

$$AU_{i,j,x} = \frac{||\{T_{i,x}\} - \{T_{j,x}\}||}{||\{T_{i,x}\}||},$$

where $\{T_{i,x}\}$ ($\{T_{j,x}\}$) is the set of relevant documents retrieved at point x in the search by methods i and j, respectively, and $\|\{T\}\|$ is the cardinality or number of elements in set T. In this case, the subtraction operator should be understood as a set difference operator. This measure is useful when combining different ranking methods to obtain the greatest degree of document coverage. If we begin with what is considered the best ranking method, the next method to be used should be the method which retrieves the most additional relevant documents, given that the first method has retrieved a set of relevant documents. This is equivalent to or better than choosing the method which does the next best job retrieving documents, which might retrieve no documents that haven't already been retrieved by the first method.

A simple, commonly used single number measure computes the average precision of searches at specific points, usually the 25%, 50%, and 75% recall levels. If one simplifies the searching into two types, brief searches where the searcher desires only a few relevant documents, and exhaustive searches, where almost all the relevant literature is located, then measuring performance by computing the average at 25, 50, and 75% recall levels is intuitively appealing.

Performance when R documents have been retrieved and where there are R relevant documents in the database is referred to as *R-precision* (Witten, Moffat, & Bell, 1994). Interestingly, the precision at this point is the same as the recall.

Normalized Recall and Precision

A desirable characteristic of a measure is that it has a zero value when it is worst-case and a one when it is best-case. This can be obtained with normalized recall, which compares the recall level with the upperbounds and lowerbounds for recall.

$$recall_{norm} = 1 - \frac{\sum_{i=1}^{R} rank_i - \sum_{i=1}^{R} i}{R(N - R)},$$

where R is the number of relevant documents, N the number of documents, and $rank_i$ the rank or position at which the ith document was retrieved. The numerator for the expression is the difference between the actual and the best-case and the denominator is the difference between the best-case and the worst-case.

Similar to the normalized recall is normalized precision. Taking on a value of 1 when the precision is at its best and 0 when it is worst-case, it is computed

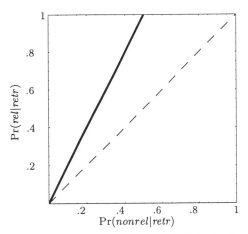

Figure 4.3.   Receiver Operating Characteristic (ROC) distribution (solid line) compared to random retrieval (dashed line).

as

$$precision_{norm} = 1 - \frac{\sum\limits_{i=1}^{R} \log \; rank_i - \sum\limits_{i=1}^{R} \log i}{\log \dfrac{N!}{(N-R)!R!}}$$

with notation as above.

EXERCISE 4.5. [*Moderate*] For the data

    r n r r n n n r n,

compute the normalized recall and normalized precision.

Swets' Receiver Operating Characteristic E

One form of filtering performance analysis is based on the ability of a system to separate "signals" from "noise." The retrieval of documents with signals (relevant documents) is preferred to the retrieval of documents with noise (non-relevant documents). The ability of the system to separate relevant documents from non-relevant documents is measured as E and is independent of where one decides to take the measurement when examining a sequence of documents, unlike precision and recall.   The retrieval performance may be plotted as in Figure 4.3, where the solid line on the diagonal of the plot represents equal probabilities that a document is retrieved given that it is relevant and that it is retrieved given that it is non-relevant. Optimal performance here would be represented by the presence of a vertical line, which occurs when there are no non-relevant documents being retrieved while all the relevant documents are being retrieved.

Swets' E measure serves as a single number measure of text retrieval performance based on the Receiver Operating Characteristics (ROC) model (Egan, 1975; Swets, 1969). Swets' E measure may be computed as

$$E = \frac{\mu_r - \mu_n}{(\sigma_r + \sigma_n)/2},\tag{4.2}$$

where μ_r is the mean frequency for relevant documents, σ_r is the standard deviation for the frequency of relevant documents, with similar notation for non-relevant documents. It serves as a discrimination measure, determining the degree of difference between the means for the two distributions normalized. This measure addresses the performance of an ordering, unlike recall or precision, which measure performance up to a certain point in the retrieval operation.

Brookes (1968) suggests that a better form of this measure that has some statistical advantages over Swets' E is

$$S = \frac{\mu_r - \mu_n}{(\sigma_r^2 + \sigma_n^2)^{\frac{1}{2}}}.\tag{4.3}$$

This measure achieves the goals set by Swets and, because it has a known sampling distribution, it is possible to compute the statistical significance of differences between two S values.

While Swets' and Brookes' measures are justified by concerns about the power of systems to discriminate, other considerations or assumptions may lead to different measures.

The MZ based E Measure

Another E measure, which has been proposed and justified to varying degrees by Heine (1973), Van Rijsbergen (1974), and Shaw (1986), is based on desirable measurement theoretic considerations (Bollmann & Cherniavsky, 1981).

The E measure has been most simply formulated by Shaw (1995) as

$$E = 1 - \frac{2}{P^{-1} + R^{-1}},$$

combining precision (P) and recall (R) to yield a single number measure combining recall and precision values. The E measure is frequently applied at a specific point in the retrieval process, such as after retrieving 10 or 20 documents. The maximum (worst) value for E will approach 1 when precision and recall both approach 0, while the minimum value for E occurs when both precision and recall approach 1. Shaw, Burgin, and Howell (1997) find it helpful to use an F measure, computed as $1 - E$. F is the harmonic mean of the precision

Figure 4.4.   The E measure for the data in the previous figures.

and recall and has an advantage over E in that larger values are superior to smaller values, unlike E, where the reverse is true.

One may report the average for a set of E (or F) values. These individual E values may represent the performance at the same point in the retrieval or filtering process for each query, such as after receiving 10 documents, or the optimal E values for each query may be averaged. The *optimal E* for a query is the E value representing the best performance at any point during the retrieval process. In some cases, this may be when few documents have been retrieved, while in other cases the optimal E is found at a high recall level after retrieving a large number of documents.

EXERCISE 4.6. [*Easy*] What is the MZ based E value when half the relevant documents have been retrieved (and one quarter of all documents) and half the documents retrieved are relevant?

EXERCISE 4.7. [*Moderate*] What is the relationship between the F measure and R-precision?

EXERCISE 4.8. [*Moderate*] What is the relationship between the F measure and the expected value of Dice's coefficient (Equation 3.3)? How might the expected value of Dice's coefficient be used to help us understand the F measure?

4.3   AVERAGE SEARCH LENGTH (ASL)

One measure of text retrieval performance is how many documents are examined in getting to the average position of a relevant document in the ordered list of documents.

Definition 4.1 (ASL) *The Average Search Length (ASL) is the expected position of a relevant document (EPRD) in the ranked list of documents. Doc-*

uments assigned equal weights for ranking purposes are grouped together and their average position is used when calculating ASL.

In the case where the distribution of relevant documents is symmetric around the mean, the mean position of the relevant documents will equal the median position of the relevant documents, and this position will be at the fifty percent recall point, that is, the ASL is how many documents would be examined in getting to the fifty-percent recall level. If the document set had a relevant document, followed by a non-relevant document, followed by a relevant document, the average position would be in the middle and the ASL would be 2. If instead of one non-relevant document in the middle, there were two, then the ASL would increase to 2.5.

Note that if there five relevant documents and all of them were at the beginning of a set of ranked documents of any size, the ASL would be 3, that is, one half of the number of documents, plus 1/2. Similarly, if there were four relevant documents and all were at the beginning of the ranking, the ASL would be $4/2 + 1/2 = 2.5$, which is the mid-position for the four documents.

The ASL measure has several advantages over other measures. It is

- a single number measure,

- easily interpreted by the end user, and

- easily estimated analytically.

The measure is readily interpreted by the end user because it is measured in units of *documents*. When there are N documents, 1 is the lowest (best) value for the ASL and N is the highest (worst) value. For example, the ASL might be 70 documents, that is, an average of 70 documents will be retrieved in retrieving the relevant document in the average position for a relevant document. This is better for the user than having an ASL of 80 but is worse than an ASL of 67.

These numbers have a direct, measurable impact on the end user, measured in terms of the increased time it takes to retrieve relevant documents when the ASL is increased. The ASL doesn't mask poor performance, as can precision-recall curves. Stating that the average position of a relevant document is 5 is far different from saying that it is 100, and this difference is more clearly made when using the ASL than by using some other measures or combinations of measures, such as the precision-recall curve. The ASL, like all other measures based on an arithmetic mean, is sensitive to outliers (Dunlop, 1997).

The author frequently recalls using a database of approximately one thousand documents to test a ranking technique and finding significant improvements using precision-recall curves to measure performance. When the results were provided in the form of ASL measures, the ASL was several hundred documents and the improvements were on the orders of tens of documents (Losee, 1994a). The ASL would have been in the tens of thousands to the millions if the technique was used on many of the Internet indexing services. This relatively poor level of performance was masked by the precision-recall graph, although both the precision-recall data and the ASL did show the improvement due to better retrieval techniques. What the ASL captures is how bad the base-line performance really is: the reader is encouraged to note how close performance is to random in some of the graphs in this book, or how far it is from the best-case performance, and what this would mean for the user of a database of millions of documents.

In addition, the cost to the system increases directly as the ASL increases and as the amount of time the searcher uses system resources increases. Because of these direct applications of the ASL, as well as the ease with which it may be understood, the ASL is a useful measure of text filtering and retrieval performance.

Less obvious at this point is the ease with which the ASL may be computed from probabilistic parameters of the distribution of features in all documents and the distribution of features in relevant documents. When the ASL is used later in the book, performance may be directly computed from the characteristics of the documents and the queries. Once a general model is developed at the end of Chapter 6, we can relax some of the requirements placed on the ASL, such as binary relevance. Also, measures may be based on all relevant documents retrieved up to a certain point in the search and not on all relevant documents in the database.

Filtering Performance with a Cutoff

In applications where there is a cutoff point separating retrieved from non-retrieved documents, such as occurs in filtering systems or Boolean retrieval systems, one may compute the ASL for all or some of the ranked documents. Alternately, one may choose a variant to the ASL by instead locating the expected position of a properly or *appropriately classified document*, rather than the expected position of a relevant document. Documents that are relevant that are retrieved are considered *appropriate*, as are those that are not relevant and are not retrieved. All other documents are considered *inappropriate*. By determining the *expected position of an appropriate document* (EPAD), we can compute a single number measure of the performance, taking into consideration the cutoff. As with the ASL, smaller values for the EPAD indicate better performance than larger values.

EXERCISE 4.9. [*Moderate*] Assume that relevant (r) and non-relevant (n) documents are retrieved in the following order (from left to right):

r r n r n r.

The first three have the same document profile, and the last three have the same profile. What is the ASL for this data retrieval operation?

EXERCISE 4.10. [*Moderate*] Under what conditions will the ASL and the EPAD have values that are the same or approach each other?

ESL

Cooper (Cooper, 1968; Dunlop, 1997) proposed a measure of retrieval performance similar to the ASL that counted the number of non-relevant documents retrieved in the process of retrieving a chosen relevant document. Consistent with economic concerns, the *expected search length* (ESL) treats the act of retrieving a non-relevant document as having a cost, with the cost being the same for each non-relevant document. Retrieving a relevant document is treated as having no cost or loss. Counting the number of non-relevant documents retrieved in retrieving a chosen relevant document measures the cost of a retrieval action (Cooper, 1973). The mean ESL may be computed for a set of queries and retrieved sets of documents. The cost of a particular retrieval mechanism thus will be proportional to the mean ESL associated with the mechanism. As one decreases the ESL, one decreases the cost to the searcher of the retrieval action.

One can scale measures such as ASL or ESL so that they reflect the degree to which they vary from optimal retrieval or random retrieval. For example, the *expected search length reduction factor* (ESLR), the reduction in expected search length from random performance, may be computed as a normalized measure. Given the ESL for a random ordering of documents, the ESLR may be computed as the difference between the ESL for random ordering and the ESL, divided by the ESL for random ordering. This will be at its maximum of 1 when the ESL is 0. The ESLR is 0 when the ESL for the method in question is the same as the ESL for random ordering.

EXERCISE 4.11. [*Moderate*] Under what conditions would ESL and ASL have the same value?

4.4   SUMMARY

There are a variety of information retrieval and filtering measures useful in evaluating system performance. Most of these measures are used largely to describe the output of systems, based on the ordering of relevant and non-relevant documents. Some single number measures have attractive intellectual foundations, such as measure theory or economics. The ASL can serve as an easy-to-understand single number measure and can also be predicted, as will be seen in the next two chapters.

5 THE QUALITY OF A RANKING METHOD

There is [a]... great cause of the little advancement of the sciences, which is this: it is impossible to advance properly in the course when the goal is not properly fixed.

—Francis Bacon, 1620

5.1 INTRODUCTION

The most commonly used performance measures represent the characteristics of a search up to a certain point in the retrieval process. These measures include precision, recall, and Van Rijsbergen's and Shaw's E measure. Swets' E measure, on the other hand, addresses the performance of a particular ranking based on the characteristics of the distributions for relevant and non-relevant documents. It is therefore independent of the stopping point chosen in a particular search. The relationship between ordered sequences, such as lists of ranked documents, may be studied using a number of techniques (Levenstein, 1966; Myers & Miller, 1988), ranging from sophisticated cost-based methods to counting the number of swaps necessary in sorting one sequence so that it becomes the other sequence. How best to study the relationships between sequences is an open question.

Below, we take a different approach, *studying ranking formulae themselves* to determine when and how they act optimally. This form of measure may be used to compare retrieval performance for different ranking methods under a variety of different conditions. Most importantly for our purposes, it may also be used in the analytic determination of the ASL for a given set of conditions.

Below, we develop a measure of the degree of optimality of a ranking procedure, whether a simple ranking formula or a complex formula representing a mixture of other, simpler formulae. This technique is then applied to a number of term weighting systems for the case of single terms. By examining the circumstances under which a ranking algorithm is optimal, combined with considering how knowledge, ignorance, and misinformation affect performance, we will be able to make general statements about filtering performance. When these techniques are combined to produce methods for computing the ASL for optimal and sub-optimal ranking in Chapter 6, numeric performance figures are generated without using the simulations that dominate traditional information retrieval research. In addition, using these measures and methods will lead to the additional understanding of *why* one method is better than another.

Optimal Rankings

Assume that the set of available documents $\mathbf{v}$ is weakly ordered by a ranking algorithm $\mathcal{R}_x$ and there are $N = |\mathbf{v}|$ elements in the set of documents $\mathbf{v}$. The ranking $\mathcal{R}_x$ produces an ordering of the documents $\mathbf{v}_y = d_1 \succeq d_2 \succeq \cdots \succeq d_N$, that is, d_1 is weakly ordered before d_2 which is weakly ordered before d_3, and so forth to d_N. When studying the ranking of a set of documents and the resulting performance, it may be necessary to compute $\Pr(\mathbf{v}_y|\mathcal{R}_x)$ the probability that one will obtain a particular document ordering $\mathbf{v}_y$ when ranking procedure $\mathcal{R}_x$ is used.

Definition 5.1 (Probability of the optimal ranking) *One ranked set of documents is the best, or optimal, denoted as $\mathbf{v}_o$, and the probability that we will have this when the optimal ranking ($\mathcal{R}_o$) procedure is used is* $\Pr(\mathbf{v}_o|\mathcal{R}_o) = 1$.

The probability $\Pr(\mathbf{v}_y)$ may be understood as the proportion of the series of queries meeting the assumptions of the model that have ordering y, given the probabilities associated with the query. The probability may represent the subjective probability that an individual believes that the ranking will be as given.

Measuring the Degree of Optimality

Sometimes it may be helpful to measure the degree of optimality for a ranking procedure $\mathcal{R}$, the probability the ranking is the same as that provided by optimal ranking ($\mathcal{R}_o$). This, combined with other performance factors, can allow us to compute expected performance in the single term case.

Definition 5.2 (Degree of optimality $\mathcal{Q}$) *The probability that ranking methods $\mathcal{R}_x$ and $\mathcal{R}_o$ produce the same ranking, computed over the set of rankings,* *is*

$$Q_x = \Pr(\mathcal{R}_o, \mathcal{R}_x) = \sum_{i \in Perm(\mathbf{v})} \Pr(\mathbf{v}_i, \mathcal{R}_x, \mathcal{R}_o), \qquad (5.1)$$

where $Perm(\mathbf{v})$ is the set of distinct weak orderings possible, rearrangements of the document profiles in vector $\mathbf{v}$.

In the case of using a single term in queries and documents, two possible orders exist: placing documents with the feature first and the documents without the feature second, or the reverse. Note that for the case where there is more than one term or more than two possible orders, Q_x denotes the probability of obtaining document ordering x.

We consider here cases where there are documents with identical profiles. Q is normed and has the value of 1 when ranking is optimal and 0 in the worst-case. It is used to average joint probabilities over the two sets of ordered documents, where, $1 \succ 0$ and $0 \succeq 1$.

The simplest environment in which we may examine the degree of optimality exists when there is only a single term in the query. When two documents have the same profile, and all characteristics being considered by the ranking procedure are the same, any two deterministic methods will provide the same ranking. When there is a difference between the documents, the ranking procedure selected determines which ranking is to be preferred. To compute Q for this single term case, consider the probability that the ranking of documents for binary feature d is the same for two ranking functions (in this case the ranking function $\mathcal{R}_x$ and the optimal ranking function $\mathcal{R}_o$), in the case where the document with a 1 is ranked ahead of or the same as a document with profile 0 (denoted here as $1 \succeq 0$) and the case where a document with a 0 is ranked ahead of a document with a 1 (denoted here as $1 \prec 0$). For purposes of notational simplicity, we consider the documents to have been re-parameterized so that the feature value 1 is assigned to the feature characteristic that produces the highest discrimination value under optimal ranking.

Theorem 5.1 (Degree of optimality for a term) *When there is a single term in the query, the probability that ranking method $\mathcal{R}_x$ is optimal is,*

$$Q_x = \Pr(\mathcal{R}_x, \mathcal{R}_o, 1 \succ 0) + \Pr(\mathcal{R}_x, \mathcal{R}_o, 1 \preceq 0), \qquad (5.2)$$

the probability that the ranking method in question and the optimal ranking have the same ordering.

When there is only a single, binary feature being considered (which we assume below unless specified otherwise), there are only two orderings possible: documents with the feature ordered before those without the feature ($1 \succ 0$) and documents without the feature ordered before or the same as documents with the feature ($1 \preceq 0$). Because Q_x is the degree of overlap between ranking $\mathcal{R}_x$ and the optimal ranking $\mathcal{R}_o$, $Q_x = \Pr(\mathcal{R}_o, \mathcal{R}_x)$.

Table 5.1.   Computing the probabilities for Q for 4 single term queries and 6 documents. Here, $p = \Pr(d = 1|rel)$ and $t = \Pr(d = 1)$.

Document	Relevance for Queries			
Profiles	1	2	3	4
0	N	N	N	N
0	Y	Y	N	N
0	N	Y	Y	Y
1	N	N	Y	N
1	Y	Y	Y	N
1	Y	Y	Y	N
p:	2/3	1/2	3/4	0
t:	1/2	1/2	1/2	1/2

Parameter Orderings

Computing Q for different ranking methods requires in some cases that estimates be made of the probability that one parameter exceeds or is less than a second parameter. The probability that one parameter is greater than another, or that it is within a certain range, may be computed from historical data. For example, the data in Table 5.1 has $p \geq t$ for three of the four queries, thus $\Pr(p \geq t) = 3/4$, while $p < .5$ one quarter of the time (query 4,) yielding a probability $\Pr(p < .5) = 1/4$. Given historical data, and knowing the formula for Q for a specific ranking method, one can determine which ranking method will perform best for a given set of parameters and thus a set of database and query pairs.

This approach to computing Q is used when one desires to historically analyze a set of executed queries. A Bayesian method can predict Q for the future, using the distribution of values for a given parameter. This model allows us to make claims about future performance, given all available current knowledge.

5.2   DEGREE OF OPTIMALITY FOR SPECIFIC MODELS

The retrieval status value for a document in which the single term has weight or discrimination value w and term frequency d is denoted as the product dw. We find that for binary term frequencies, documents with feature frequency 1 are ordered before documents with a 0, that is, $(1 \succ 0)$, if and only if $1w > 0w$, that is, when w is greater than zero. One can similarly compute the probability that $(1 \preceq 0)$ as the probability that $1w \leq 0w$, which is the probability that w is less than or equal to 0.

In several sections below, specific retrieval algorithms will be studied to determine their degree of optimality. These results can be used in several different ways. One is as a criteria for determining when ranking is optimal. We will find, for example, that IDF ranking, for a single term, produces optimal ranking when $p > t$. One can also compute Q as representing the average degree of optimality, given a large number of queries. Thus, the IDF weighting is optimal in the those cases where $p > t$ and the probability it is optimal is $\Pr(p > t)$. A third use of Q addresses our subjective belief about the degree of optimality for a single query, given available knowledge about the parameters of the model. For example, a professional, based on their own expertise, may believe that for an individual search using the IDF weighting, the probability that the search will be optimal is $3/4$, that is, that p will be greater than t about $3/4$ of the time for queries such as the one in question.

Best-case Ranking

The optimal ranking of documents is obtained by ordering documents by decreasing expected precision, which is the same as ranking them by the probability that they are relevant, $\Pr(rel|d)$. Ranking documents by their probability of relevance, given that they have certain features, places documents with higher probabilities of relevance ahead of documents with lower probabilities of relevance. We assume for the next several sections that the probabilities being considered are point probabilities.

This best-case ranking is the same as the ranking obtained with

$$\Pr(rel|d) = \frac{\Pr(d|rel)\Pr(rel)}{\Pr(d)}.$$

Because the $\Pr(rel)$ component is constant for each document, it may be dropped without affecting ranking; thus, ranking documents by $\Pr(d|rel)/\Pr(d)$ produces the same ranking. Given binary features, the weight for a feature is

$$w = \log\left(\frac{p/(1-p)}{t/(1-t)}\right), \tag{5.3}$$

where $p = \Pr(d = 1|rel)$ and $t = \Pr(d = 1)$.

To apply Equation 5.1 we must compute the probability that a document with a feature value of 1 is ranked ahead of or equal to a feature with value 0:

$$1w > 0w$$
$$\log\left(\frac{p/(1-p)}{t/(1-t)}\right) > 0. \tag{5.4}$$

This is the case only when $p > t$. When this condition is met, the number in the left hand side of Equation 5.4 is positive.

Similarly, we compute the probability that a document with a feature value of 0 is ranked ahead of a feature with value 1 as

$$1w \leq 0w$$

$$\log \left(\frac{p/(1-p)}{t/(1-t)} \right) \leq 0.$$

We find that this is the case when $p \leq t$.

The probability that optimal ranking is obtained when the ranking function $\mathcal{R}$ is $\mathcal{R}_o$ is

$$
\begin{aligned}
\mathcal{Q}_B(p,t) &= \Pr(p > t, p > t) + \Pr(p \leq t, p \leq t) && (5.5) \\
&= 1.
\end{aligned}
$$

Worst-case Performance

The worst-case ordering is that obtained when documents with different term frequencies are always ranked the opposite by $\mathcal{R}$ than by the optimal ranking $\mathcal{R}_o$. This requires that we compute the following probabilities: $\Pr(\mathcal{R}_{o,1 \succ 0}, \mathcal{R}_{x,1 \preceq 0})$ and $\Pr(\mathcal{R}_{o,1 \preceq 0}, \mathcal{R}_{x,1 \succ 0})$.

Using the results derived in Equation 5.5 for the probability of optimality for the best case ranking, this can be seen to be

$$
\begin{aligned}
\mathcal{Q}_W(p,t) &= \Pr(p > t, p \leq t) + \Pr(p \leq t, p > t) && (5.6) \\
&= 0.
\end{aligned}
$$

This problem may be viewed as suggesting the ranking of documents by $1 - \Pr(rel|d) = \neg \Pr(rel|d)$, where $\Pr(rel|d)$ is the optimal weighting, if we wish to achieve worst-case performance. The probabilities obtained in Equation 5.6 may be treated as

$$
\begin{aligned}
\mathcal{Q}_W(p,t) &= \Pr(p > t, \neg(p > t)) + \Pr(p \leq t, \neg(p \leq t)) \\
&= 0.
\end{aligned}
$$

For notational purposes, the joint probabilities used in describing the $\mathcal{Q}$ values will always list the optimal condition first, followed by the condition for the model whose probability of optimality is being computed.

Random Ranking

The random ranking of documents could be expected to give optimal ranking half of the time and worst-case ranking the other half. Using Equation 5.2, one can compute

$$\mathcal{Q}_R(p,t) = \Pr\left(p > t, \frac{p > t}{2}\right) + \Pr\left(p \leq t, \frac{p \leq t}{2}\right) = \frac{1}{2} \qquad (5.7)$$

where the fractions $\Pr((p > t)/2)$ and $\Pr((p \leq t)/2)$ represent the portion (one half) of each probabilistic distribution that is to be combined in the joint probability. Exactly which documents are used (from which this random half is taken) is not important for purposes here.

IDF

Inverse document frequency (IDF) weighting is an effective feature weighting system when no information about relevant documents is available, other than the fact that a term is or is not in the query (Sparck Jones, 1972; Yu & Salton, 1977; Croft & Harper, 1979). Using as a weight

$$w = -\log t,$$

we find that the weight increases as t decreases.

One finds that $1 \succ 0$ if and only if $-\log t > 0$, that is, when $t > 0$, which is the case for all valid probabilities, all values of t. For the other ordering, $1 \precsim 0$ is found if and only if $-\log t \leq 0$, which never occurs, for any valid probability t.

Equation 5.2 may be used to compute

$$
\begin{aligned}
\mathcal{Q}_I(p,t) &= \Pr(p > t, t > 0) + \Pr(p \leq t, t \leq 0) \qquad (5.8) \\
&= \Pr(p > t).
\end{aligned}
$$

P-Weighting

For illustrative purposes, let us consider weighting documents by

$$d \log \frac{p}{1-p}.$$

We will call this *p-weighting* and propose it for illustrative purposes. Documents are ranked so that $1 \succ 0$ when $\log((p/(1-p))/(t/(1-t))) > 0$ (from the best-case ordering) and $\log(p/(1-p)) > 0$ (from the p-weighting). These inequalities are met when $p > t$ and $p > .5$. The ordering $1 \precsim 0$ is obtained when $\log((p/(1-p))/(t/(1-t))) \leq 0$ and $\log(p/(1-p)) \leq 0$, occurring when $p \leq t$ and $p \leq .5$. We can thus derive the $\mathcal{Q}$ value for p-weighting as

$$\mathcal{Q}_P = \Pr(p > t, p > .5) + \Pr(p \leq t, p \leq .5). \qquad (5.9)$$

Example.   Consider the set of $\{p, t\}$ pairs $\{.8, .2\}, \{.6, .3\}, \{.4, .45\},$ and $\{.6, .7\}$, characterizing 4 different searches. We note that for the first two parameter pairs, both $p > t$ and $p > .5$. For the third pair, $p \leq t$ and $p \leq .5$. The last $\{p, t\}$ pair clearly doesn't fall into either of these two categories. Since 3 of the 4 pairs would be included in one of the two probabilities summing to produce $\mathcal{Q}_P$ in Equation 5.9, we compute $\mathcal{Q}_P = 3/4$.

Decision Theoretic Weighting

The weighting provided by Equation 3.12 is possibly the most widely supported weighting for binary independent features. We find for ordering $1 \succ 0$ that

$$\log\left(\frac{p/(1-p)}{q/(1-q)}\right) > 0$$

only when $p > q$.

We find for ordering $1 \preceq 0$ that

$$\log\left(\frac{p/(1-p)}{q/(1-q)}\right) \leq 0$$

only when $p \leq q$.

We can thus calculate

$$\mathcal{Q}_D(p,t,q) = \Pr(p > t, p > q) + \Pr(p \leq t, p \leq q). \qquad (5.10)$$

The variable t will always be a value between p and q, that is, either $p \geq t \geq q$ or $p \leq t \leq q$. The $\mathcal{Q}_D(p,t,q)$ value can be interpreted as 1 minus the probability of being in the gap between t and q. This gap will be very small for large datasets where almost all documents are non-relevant and the value of t approaches the value of q.

Coordination Level Matching

Coordination level matching (Van Rijsbergen, 1986; Losee, 1987) suggests the retrieval of documents arranged in decreasing order of the number of terms in common with the query, referred to as the *coordination level.* Used in the early Cranfield indexing and retrieval tests as a weighting method, it computes the document weight as the summation of the binary term frequency times a weight constant for all terms, or

$$\sum_{i=1}^{n} d_i c. \qquad (5.11)$$

When this is implemented as a series of Boolean expressions, first retrieving documents with all terms, then retrieving documents with all but one term, and so on until documents with no terms are retrieved, it is referred to as quorum level searching (Salton, Wong, & Yu, 1976).

A document with frequency 1 is retrieved ahead of a document with feature frequency 0 if, and only if, $c > 0$, that is, if c is positive. Similarly, we have the ordering $1 \preceq 0$ if and only if $c \leq 0$. If we assume that c is always positive for

Table 5.2.   Comparing ranking methods.

$\mathcal{R}$		$\mathcal{Q}$
$\mathcal{R}_o$ = Best-case	$\mathcal{Q}_B =$	$\Pr(p > t, p > t) + \Pr(p \leq t, p \leq t)$
	$=$	1
Random	$\mathcal{Q}_R =$	$(\Pr(p > t) + \Pr(p \leq t))/2$
	$=$	$1/2$
Worst-case	$\mathcal{Q}_W =$	$\Pr(p > t, p \leq t) + \Pr(p \leq t, p > t)$
	$=$	0
Dec Theo	$\mathcal{Q}_D =$	$\Pr(p > t, p > q) + \Pr(p \leq t, p \leq q)$
	$=$	$1 - (\Pr(p > max(t, q)) + \Pr(p \leq min(t, q)))$
IDF	$\mathcal{Q}_I =$	$\Pr(p > t, t > 0) + \Pr(p \leq t, t \leq 0)$
	$=$	$\Pr(p > t)$
CLM	$\mathcal{Q}_C =$	$\Pr(p > t)$

query terms,

$$\begin{aligned} \mathcal{Q}_C(p, t) &= \Pr(p > t, c > 0) + \Pr(p \leq t, c \leq 0) & (5.12) \\ &= \Pr(p > t). \end{aligned}$$

EXERCISE 5.1. [*Moderate*] How does one compute $\mathcal{Q}_D$, the degree of optimality under the decision theoretic model (Equation 5.10) assuming that perfect knowledge about p is beta distributed and that q is beta distributed?

Comparing Ranking Methods

Table 5.2 summarizes the $\mathcal{Q}$ values for several filtering models. Clearly the optimal ranking is best, with $\mathcal{Q}_B = 1$, and the worst is $\mathcal{Q}_W = 0$. In cases where the query term is a positive discriminator, which it usually is, IDF, CLM, and the decision theoretic models perform better than random but worse than optimal retrieval. One can see that as the difference between q and t approaches 0, $\mathcal{Q}_D$ approaches 1.

EXERCISE 5.2. [*Moderate*] What would be $\mathcal{Q}$ for the ranking of documents by $d \log q/(1 - q)$? How does this compare to what is obtained for the IDF measure?

EXERCISE 5.3. [*Moderate*] Under what conditions would $\mathcal{Q}_D$ be lowest? What value would $\mathcal{Q}_D$ approach under these circumstances?

EXERCISE 5.4. [*Moderate*] Given the ranking formula $d \log(p - t)$, what is this ranking method's $\mathcal{Q}$ value?

5.3   RELEVANCE FEEDBACK

Changing the values of parameter estimates changes the degree of optimality for a ranking procedure. Incorporating relevance feedback essentially changes the operating characteristics of the ranking algorithm $\mathcal{R}$ by modifying the parameters of the prior distribution describing our knowledge about the parameters describing term frequencies in documents.

Relevance feedback is user supplied information about the relevance of documents retrieved by the system. Some assumptions are made below, including that relevance is binary, that is, a document is relevant or it is not relevant, and that the judgements that are provided are binary, that is, they are not probabilistic judgements, such as believing that there is a thirty percent chance that a document is relevant and, conversely, a seventy percent chance that it is not relevant. The requirement that relevance and relevance judgements are binary is dropped near the end of this chapter.

To estimate $\mathcal{Q}$ it is necessary to estimate probabilities for p, t, and q. While these probabilities may be point probabilities such that either $p \geq t \geq q$ or the reverse holds, the probabilities may also be estimated as distributions. In some situations, our knowledge about these parameters for binary distributions may best be represented by beta distributions, describing the probability that the parameter p of the binary distribution has a certain value. We often can treat t as a point probability, since we can always know precisely what percent of documents have a particular term. The parameters p and q are not known before hand, although for large databases we may often make the approximation $q \approx t$ and thus treat q as a point probability.

The parameter values for the beta distribution may be learned through relevance feedback provided by the searcher. As the user provides judgements of "relevant" or "non-relevant" about retrieved documents, the parameters increase in value and the variance of the distribution decreases, increasing the accuracy of the estimate provided by the distribution.

When computing $\mathcal{Q}$ for different models, the probability that one parameter was above or below another was encountered. This may be implemented by using a form of the incomplete beta distribution to represent our knowledge about a parameter such as p:

$$\beta_x^y(a, b) = \frac{\Gamma(a + b)}{\Gamma(b)\Gamma(a)} \int_x^y p^{a-1}(1 - p)^{b-1}\, dp.$$

This allows us to compute the portion of a distribution within a certain range, the probability that one parameter, for example, is greater than another. In this case, the distribution represents the probability that the random variable lies within the range from x to y.

As an example, consider the application of limited knowledge to p-weighting where our knowledge about p is represented by a beta distribution:

$$\begin{aligned}
\mathcal{Q}_P &= \Pr(p > t, p > .5) + \Pr(p \leq t, p \leq .5) \\
&= min\left(\beta_t^1(\mu \to p, \sigma^2 \to 0), \beta_{.5}^1(a, b)\right) \\
&\quad + min\left(\beta_0^t(\mu \to p, \sigma^2 \to 0), \beta_0^{.5}(a, b)\right).
\end{aligned} \qquad (5.13)$$

We treat optimal ranking as though p were described by a beta distribution here with a mean approaching the value p ($\mu \to p$) (if p is a point probability and relevance feedback improves performance) and with the variance approaching 0 ($\sigma^2 \to 0$). This approximates a point estimate at p. The second beta distribution, with parameters a and b, describes the estimate for p held by the user of the model.

The degree of optimality $\mathcal{Q}_P$ is at its maximum, that is, p-weighting approaches optimal ranking, when the beta distribution describing the user's knowledge matches the distribution describing the p for optimal ranking. This is the case where the user has perfect knowledge.

The minimum value for $\mathcal{Q}_P$ occurs when the mean for the beta distribution describing p in the optimal model (the true value of p) is within the gap between t and .5 and the distribution peaks within this area. This peaking occurs when there is a great deal of misleading relevance feedback about the characteristics of relevant documents.

5.4 COMPARING RANKING UNDER DIFFERENT LEVELS OF KNOWLEDGE

The differences in ranking quality may be computed by examining the difference between the $\mathcal{Q}$ values. The difference between methods i and j may be computed as $\Delta \mathcal{Q}_{i,j} = \mathcal{Q}_i - \mathcal{Q}_j$.

The difference between $\mathcal{Q}$ for decision theoretic and IDF ranking may be computed as

$$\Delta \mathcal{Q}_{DI} = \Pr(p > t, p > q) + \Pr(p \leq t, p \leq q) - \Pr(p > t).$$

Remember that $\mathcal{Q}_D$ is the area in the distribution describing p outside the range from q to t, and p is described by a distribution.

If the term is a positive or neutral discriminator, then it will be the case that $p \geq t \geq q$. If the knowledge about p is described by a distribution then $\Pr(p > t)$ is the cumulation of the function (distribution) from t up to 1 while $\Pr(p > q)$ is the cumulation of the function from q up to 1. Since $t \geq q$, the joint probability $\Pr(p > t, p > q)$ is the cumulation from t up to 1. Using similar techniques, $\Pr(p \leq t, p \leq q)$ is the cumulation from the 0 point in the function up to q. Similar techniques are used when the term is a negative discriminator, that is, $p < t < q$. When the term is a positive discriminator, then $\mathcal{Q}_I$ is the area from the high end of the distribution down to t. In this case, $\Delta \mathcal{Q}_{DI}$ will be the area remaining, from below q to the low end of the distribution.

When the term is a negative discriminator, then $\mathcal{Q}_I$ is the area from the high end of the distribution down to t, which will be lower than q. Thus, the area will include both from the high end of the distribution down to q and the gap from q down to t. In this case, $\Delta \mathcal{Q}_{DI}$ will be the area below t minus the gap between q and t. If the area in the gap is larger than the area below t, then one would expect IDF weighting to outperform decision theoretic weighting, i.e., $\Delta \mathcal{Q}_{DI} < 0$. If the distribution is concentrated below t, however, the area under

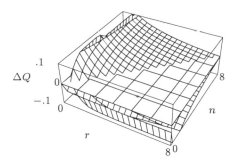

Figure 5.1.   The difference in $\mathcal{Q}$ values when using the Shaw minus the Robertson-Sparck-Jones estimates. Positive values occur in the region where Shaw's estimate is superior (the smaller grid squares are above the large squared zero plane,) while negative values are regions where RSJ is superior (the larger grid squares are above the small squares). Shaw's method is superior in the region where the number of non-relevant documents retrieved exceeds the number of relevant documents retrieved.

the distribution below t will be larger than that in the gap and decision theoretic methods will be expected to outperform IDF methods, i.e., $\Delta\mathcal{Q}_{DI} > 0$.

Example: Using $\mathcal{Q}$ to Evaluate Parameter Estimates

One application of $\mathcal{Q}$ is in the evaluation of parameter estimation techniques. Robertson and Sparck-Jones (1976) have suggested that a suitable estimate for p values when no other information is available is

$$p = \frac{r + \frac{1}{2}}{r + n + 1},$$
(5.14)

where r is the number of relevant documents retrieved with the term and n is the number of relevant documents retrieved without the term. This form of estimation has a long statistical history in a variety of applications. It can be seen as consistent with LaPlace's principle of succession (Equation 2.15) and the Bayesian models developed in Chapter 2 and 3, where the 1/2 and 1 represent parameters consistent with non-informative prior information.

Shaw (1995) has suggested a second initial estimate for p. For most values, $p = r/(n + r)$. When r would be 1, p is computed as $p = 1 - 1/N^2$ where N may be the number of documents in the database. When r would be 0, p is set to $1/N^2$.

These specific formulae for estimating parameters may be compared by computing the $\mathcal{Q}$ value for each one and then comparing them. This is done in Figure 5.1 for a range of r and n values. Here, N is set to 100.

This analysis assumes that the prior information about p is beta distributed and that ordering is otherwise optimal. In many situations these assumptions are not met, but making simplifying assumptions allows us to roughly compare the two estimation techniques. If these assumptions hold, it is clear from the figure that the Shaw method is superior as an initial estimate over one range of data values and the Robertson-Sparck-Jones method is better in another region. This is not to argue that one of these estimates is universally or conclusively better or worse than another method. Instead, we have suggested that a problem such as this can be addressed analytically. However, there are still assumptions in the analysis that need to be evaluated through data collection and improved modeling.

EXERCISE 5.5. [*Research Problem*] Under what conditions is the RSJ (Robertson & Sparck Jones, 1976) method for estimating parameters, Equation 5.14, better than random?

5.5   A GENERAL MODEL OF RANKING PERFORMANCE

A more general model of Q may be developed beginning with the work above. We denote the distribution describing perfect knowledge about p as $g(x)$, where x is a probabilistic parameter such as p.

Parameter p as a Point Probability

We may define Q when our knowledge of p is treated as a point probability, as:

$$Q = \Pr(\mu_r > \mu, X) + \Pr(\mu_r \leq \mu, Y), \qquad (5.15)$$

where X and Y represent ranking specific conditions, and μ, μ_r, and μ_n represent the expected frequency for all documents, relevant documents, and non-relevant documents respectively. Under the binary model, $\mu = t$, $\mu_r = p$, and $\mu_n = q$. As an example of ranking specific conditions, under best-case ranking $X = \mu_r > \mu$ and $Y = \mu_r \leq \mu$. For worst-case ranking, $X = \mu_r < \mu$ and $Y = \mu_r \geq \mu$.

Parameter p is Described by a Distribution

When our knowledge about p is perfect and is represented by a distribution g rather than a point probability p, then

$$
\begin{aligned}
Q &= min\left(\int_\mu^\infty g(z)\,dz, \int_x^\infty g(z)\,dz\right) \\
&\quad + min\left(\int_0^\mu g(z)\,dz, \int_0^y g(z)\,dz\right) \\
&= 1 - \int_{min(\mu,x)}^{max(\mu,y)} g(z)\,dz, \qquad (5.16)
\end{aligned}
$$

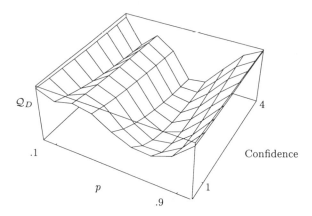

Figure 5.2. $\mathcal{Q}_D$ as p varies over the range of possible values and as the amount of knowledge increases (represented by an increased value on the bottom right axis). Confidence represents the conceptual number of documents retrieved.

where x and y represent values specific to the model being considered. For example, the optimal (best-case) binary retrieval model may be implemented in the general model of $\mathcal{Q}$ by setting g to the beta distribution, with $x = \mu$ and $y = \mu$, while with the decision theoretic model, $x = \mu_n$ and $y = \mu_n$ $(\mu_n = q)$.

An example of the application of this method to computing $\mathcal{Q}_D$ is given in Figure 5.2. As the amount of knowledge about p increases, $\mathcal{Q}_D$ varies. We set $q = .3$, and t is varied to be consistent with the corresponding value for p.

Inaccurate or Incomplete Knowledge about p

When our knowledge about p is imperfect and is represented by a distribution $h(x)$, and p is exactly distributed as described by the distribution $g(x)$, then we may compute:

$$
\mathcal{Q} = min\left(\int_{\mu}^{\infty} g(z)\, dz, \int_{x}^{\infty} h(z)\, dz\right)
$$
$$
+ min\left(\int_{0}^{\mu} g(z)\, dz, \int_{0}^{y} h(z)\, dz\right). \tag{5.17}
$$

This can be useful when we wish to evaluate the effect of assuming one distribution about p, $h(x)$, when another distribution is the correct one, $g(x)$. When $g(x) = h(x)$ for all x, then Equation 5.17 will be equivalent to Equation 5.16.

The beta distribution is often used to model binary parameters such as p but it is usually conceded that this is a simplifying approach. When the true distribution is an irregular distribution, perhaps not being easily described by any of the traditional parametric distributions, $g(x)$ would be this empirical distribution and $h(x)$ would be the estimate provided by a simplifying distribution used by the model, such as a beta or normal distribution.

Continuous Relevance

Using the methods developed earlier, a continuous form of relevance may be incorporated into the analytic model of text filtering performance. We assume that the user wants to retrieve documents with a relevance value of x or greater. Instead of using a function $g(z)$ to describe our knowledge about the distribution of p, we use the function $g(z, r)$ to describe knowledge about p_x, where r is a relevance value that has as its lowest value x and as its highest value 1, $0 \leq x \leq 1$. When one is computing the component $\int_\mu^\infty g(z)\, dz$ in Equation 5.17 (with a similar analysis for the other integrals over $g(z)$), continuous relevance may be incorporated through the use of expressions such as

$$\int_x^1 \frac{\Pr(r)}{\int_x^1 \Pr(r')\, dr'} \int_\mu^\infty g(z, r)\, dz\, dr.$$

The computation is similar to Equation 5.17 but the integral incorporating $g()$ function is averaged over the relevance values (r) from x up to 1. Similar modifications can be made to other formulae computing Q to make them consistent with the assumption of continuous relevance. Full development of this model is left as an exercise below.

EXERCISE 5.6. [*Moderate*] Design an experiment to determine whether the mean term frequencies for the different relevance values are linear with regard to the level of relevance, or whether the nature of the relationship between relevance and term frequencies is non-linear.

EXERCISE 5.7. [*Research Problem*] Assume that term frequencies are binary and that ranking uses decision theoretic weighting. If underlying relevance is continuous but it is treated as binary, what is the probability that ranking is optimal? What would this be generally for any weighting formulae, not just for decision theoretic weighting?

EXERCISE 5.8. [*Research Problem*] How would one compute Q for the situation where term frequencies are Poisson distributed but those terms with frequency greater than 0 are treated as having the binary frequency 1 and those with term frequency 0 are treated as having binary term frequency 0?

5.6 EXISTENCE AND CONSTRUCTION OF RANKING PROCEDURES

While earlier developments started from specific document ranking systems and moved toward a more general model of Q, generality can be obtained by beginning with Q and moving toward a ranking procedure, constructing at least one ranking procedure for every Q.

Theorem 5.2 (Existence of ranking procedures for any $\mathcal{Q}$) *There is at least one single term ranking procedure that has a degree of quality $\mathcal{Q}$, where $0 \leq \mathcal{Q} \leq 1$. Thus, there are an infinite number of ranking procedures, given the infinite number of values between 0 and 1.*

Let us assume that p is distributed as described by continuous distribution $g(z)$. The probability that a ranking is optimal may be computed when $g(z)$ describes the distribution of knowledge about the distribution of term frequencies in relevant documents as Equation 5.16:

$$\mathcal{Q} = 1 - \int_{min(\mu,x)}^{max(\mu,y)} g(z)\,dz.$$

When x approaches the low feature frequency value (i.e., 0 for the binary term distribution) and y approaches the high feature frequency value (i.e. 1 for the binary term distribution), we find that $\mathcal{Q}$ approaches 0. Conversely, when x and y approach μ then $\mathcal{Q}$ approaches 1. Clearly, by choosing appropriate values for x and y, one can obtain any desired value for $\mathcal{Q}$

Algorithm to construct a single-term ranking algorithm. A single binary term ranking procedure parameterized to produce $\mathcal{Q}$ (assuming it represents a positive discriminator) as $\mathcal{Q} = \Pr(p > t, a > b) + \Pr(p \leq t, a \leq b)$ may be produced with the weighting

$$w = \log \frac{a/(1-a)}{b/(1-b)}, \tag{5.18}$$

where $0 \leq a \leq 1$ and $0 \leq b \leq 1$. We assume that a is described by a distribution, as is p. Arbitrarily set point value b to t, and assume that the term is parameterized to be a positive discriminator. Assume that $\Pr(a > b) = 1 - \Pr(a \leq b)$ and that $\Pr(p > t) = 1 - \Pr(p \leq t)$. When $\mathcal{Q}$ is better than random, set distribution variable a numerically so that

$$\Pr(a > b) = \Pr(p > t) - \mathcal{Q}. \tag{5.19}$$

Note that

$$\Pr(a \leq b) \geq \Pr(p \leq t). \tag{5.20}$$

Consider the case where the mean estimate for a is lower than for p, leading to sub-optimality and $\mathcal{Q} < 1$. In this case, the condition in Equation 5.20 is met for the distributions considered here. That portion of $\mathcal{Q}$ computed from components less that t is thus $\Pr(p \leq t)$. Because of Equations 5.19 and 5.20, $\mathcal{Q} = \Pr(a > t) + \Pr(p \leq t)$. Similar procedures can be constructed for other point or distributional assumptions about $\mathcal{Q}$, p, and t.

EXERCISE 5.9. [*Difficult*] Starting with the procedure above, construct a ranking formula with $\mathcal{Q} = 3/4$. Using a method of your own design, produce a ranking formula with $Q = 3/4$. How do they differ? Why might one choose one over the other for a practical system?

5.7 SUMMARY

The ranking performance of text filtering and retrieval systems can be estimated from the parameters of term frequency distributions. The measures considered in this chapter may be used to either describe (after the fact) or to predict (before the fact) some of the characteristics and causes of the performance of document ranking systems. The use of these measures in predicting future performance will be described in the next chapter for the relatively simple case of single terms in queries. Later chapters will then examine the more complex problems associated with predicting the performance of multiple term queries and natural language phrases.

6 PERFORMANCE WITH ONE TERM

The price of reliability is the pursuit of the utmost simplicity.
—C. A. R. Hoare, 1980 ACM Turing Award Lecture

6.1 INTRODUCTION

The performance of a retrieval system can be determined in a straightforward manner with analytic techniques if one limits oneself to using only a single term from the query. Limiting our discussion to one term allows us to fully understand many retrieval characteristics and options that are far more difficult to understand in a multi-term case.

We begin with the measure of the quality of a ranking method, Q (Equation 5.2). This is combined with the characteristics of both the best-case (optimal) and the worst-case ranking to compute the Average Search Length (ASL), the expected position of a relevant document in the ranked list of documents. Then, given the characteristics of the query, the database, and of the ranking algorithm itself, the expected performance of the system can be computed.

Documents may be presented to the user in one of two different orders. Documents with a binary query feature with frequency d may be followed by those with frequency $\bar{d} = 1 - d$, which we refer to as the *optimal ordering*. It is assumed that the term weight for d is greater than for $\bar{d}$; if this is not the case, the values may be switched (re-parameterized) so that the weight for d is greater than or equal to the weight for $\overline{d}$. Thus, we assume that the features are re-parameterized so that feature frequency d multiplied by the term weight has a higher value than feature frequency $\overline{d}$ multiplied by the term weight. This is best-case or optimal ordering. The worst-case ordering retrieves documents with feature frequency $\overline{d}$ before documents with frequency d.

Definition 6.1 (A) *The parameter A is the expected proportion of documents examined in an optimal ranking if one examines all the documents up to the document in the average position of a relevant document. It is the expected position of a relevant document, scaled from 0 to 1.*

The average location of a relevant document in the optimal ranking may be computed based on a scale from 0 to 1, a unit scale, with an A of 0 representing the average position of relevant documents being at the beginning of the search process and an A of 1 being at the end of the search process.

When viewing this in an information filtering context, where some documents are passed through the filter and others are not passed, one can interpret A as 1 minus the probability a document is retrieved when retrieving documents in ranked order up to the average position of a relevant document. Similarly, if documents aren't ranked, A may be understood as 1 minus the probability that a document will be retrieved given that only documents of the average frequency of the term in relevant documents or higher are retrieved.

Theorem 6.1 (A **for single terms**) *Given binary terms with probabilities $p = \Pr(d|rel)$ and $t = \Pr(d)$, then*

$$A = \frac{1 - p + t}{2}. \qquad (6.1)$$

The variable A is computed by noting that documents with feature frequency d are at the low end of the A spectrum and those with feature frequency $\overline{d}$ at the high end of the spectrum. The middle (average) position for each of the profiles when they are arranged in order is such that $\Pr(d)/2$ is the average position for documents with feature frequency d. The mean position is $1 - \Pr(\overline{d})/2$ for the documents with feature frequency $\overline{d}$, as in Figure 6.1. Documents of each feature frequency may be weighted by the probability of having the frequency of the feature given that they are relevant. This is the percent of relevant documents that have the particular frequency. Thus,

$$A = \Pr(d|rel)\Pr(d)/2 + \Pr(\overline{d}|rel)(1 - \Pr(\overline{d})/2). \qquad (6.2)$$

The worst-case ranking that is obtained when the ordering is the opposite, that is, the placement of documents without the feature ordered before documents with the feature $(\overline{d} \succ d)$, is

$$\overline{A} = \Pr(\overline{d}|rel)\Pr(\overline{d})/2 + \Pr(d|rel)(1 - \Pr(d)/2). \qquad (6.3)$$

Here, $\overline{A} = 1 - A$.

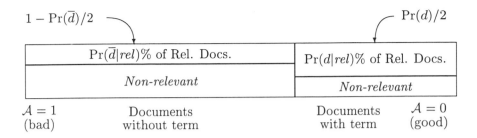

Figure 6.1.   Documents ordered for retrieval so that $1 \succ 0$.

Equation 6.2 may be simplified algebraically to:

$$\mathcal{A} = \frac{1 + \Pr(d) - \Pr(d|rel)}{2}.$$

Because the terms are binary, that is, have the values $d = 1$ or $\overline{d} = 0$, we find that

$$\mathcal{A} = \frac{1 - p + t}{2},$$

where $p = \Pr(d|rel)$ and $t = \Pr(d)$.

Example.   Consider a sequence of ten documents, the first four having the feature in question (and three of these documents being relevant) and the remaining six documents not having the feature (and one of these being relevant). We find that $\Pr(d|rel) = 3/4$, $\Pr(\overline{d}|rel) = 1/4$, $\Pr(d) = 4/10$, and $\Pr(\overline{d}) = 6/10$. Applying Equation 6.2, we compute the weighted average position of a relevant document: $(3/4)(2/10) + (1/4)(7/10) = 13/40$. Alternatively, using Equation 6.1, we find that $(1 - p + t)/2 = (1 - 3/4 + 4/10)/2 = 13/40$. It is most helpful to view this process as computing the position of the average relevant document, normed to being between 1 and 0, or it is the proportion of documents in the dataset retrieved when retrieving the conceptual document at average position of a relevant document.

EXERCISE 6.1. [*Easy*] Consider a single term query. Five relevant documents have the term and five relevant documents don't have the term. One hundred documents have the term and one thousand documents don't have the term. Compute $\mathcal{A}$.

Best and Worst-Case $\mathcal{A}$

Knowledge of the best-case and worst-case values for $\mathcal{A}$ can allow us to predict best and worst outcomes, as well as to determine how close an empirical method is to optimal. We refer to the best-case as the upper bounds for performance and the worst-case as lower bounds, although the upper bounds for performance are actually the lowest numeric value for A and ASL, with a similar direction reversal from the traditional for worst-case and lower bounds of performance representing high A and ASL values.

Theorem 6.2 (Approximate best-case A) *The best-case value for A may be approximated when $p = 1$ and $t = 0$. In this case $A = 0$.*

Corollary 6.1 (Approximate worst-case A) *The worst-case value for A may be approximated when $p = 0$ and $t = 1$. In this case $A = 1$.*

Theorem 6.3 (Exact best-case A) *The best-case value for A occurs when $p = 1$ and $t \to 0$ and the value obtained is $A = (1 - 1 + g)/2 = g/2$, where g is the generality.*

When $p = 1$, the lowest value for t becomes $g = \Pr(rel)$, the generality. Note that t will have as its low value gp, the proportion of all documents that have frequency 1 and are relevant. If we compute $A = (1 - p + gp)/2$, we find that as p increases, the $-p$ component will lower the A value faster than the gp value will raise it, for all $g < 1$. Thus, there is no intermediate value, $p < 1$, that will produce the best-case A before $p = 1$.

Corollary 6.2 (Exact worst-case A) *The worst-case value for A occurs when $p = 0$ and $t \to 1$ and $A = (1 - 0 + (1 - g))/2 = 1 - g/2$.*

As with the best-case performance, setting $p = 0$ establishes a limit on the highest value that t can have. Clearly, if no relevant documents have the feature in question, then the highest probability of a document having the feature is $1 - g$.

6.2 COMPUTING THE ASL

Being able to predict retrieval performance based on the conditions that exist at the start of the retrieval process is the cornerstone of understanding text filtering and retrieval.

Theorem 6.4 (ASL for single terms) *Given N documents and A (Equation 6.1) and Q (Equation 5.2) as defined earlier, then the Average Search Length (ASL) when using a single query term is*

$$ASL = N \left(QA + \overline{QA} \right) + 1/2.$$

To estimate the ASL it is necessary to compute the the weighted average of the A and $\overline{A}$ values. The weighting factors are the percent of orderings that are optimal and the percent of rankings that are worst-case. The probabilities or weightings used in computing the average are Q for A and $\overline{Q}$ for $\overline{A}$, where

$\overline{Q} = 1 - Q$. We may compute

$$ASL = N\Big(Q[\Pr(d|rel)\Pr(d)/2 + \Pr(\overline{d}|rel)(1 - \Pr(\overline{d})/2)]$$
$$+ \overline{Q}[\Pr(\overline{d}|rel)\Pr(\overline{d})/2 + \Pr(d|rel)(1 - \Pr(d)/2)]\Big)$$
$$+ 1/2 \qquad (6.4)$$
$$ASL = N\left(Q\mathcal{A} + \overline{Q}\overline{\mathcal{A}}\right) + 1/2. \qquad (6.5)$$

The ASL may be estimated more simply if optimal retrieval is assumed. We find that the ASL for optimal ranking, that is, where $\mathcal{R}_o$ is the ranking method used, is

$$ASL_O = N\frac{1 - p + t}{2} + \frac{1}{2}$$
$$= N\mathcal{A} + \frac{1}{2}. \qquad (6.6)$$

We may denote the ASL performance $\mathcal{P}$ given p, t, and a database of size N as

$$\mathcal{P}(p_e, p, t_e, t, q_e, q, N) = N[\mathcal{Q}(p_e, p, t_e, t, q_e, q)\mathcal{A}(p, t, q)$$
$$+ \overline{\mathcal{Q}}(p_e, p, t_e, t, q_e, q)\overline{\mathcal{A}}(p, t, q)] + \frac{1}{2} \quad (6.7)$$

where the parameters subscripted with e are estimates and those without subscripts are the correct or actual values.

The theorems here assume that $Q = 1$.

Theorem 6.5 (Approximate best-case ASL) *If $p = 1$ and $t = 0$, then $\mathcal{A} = 0$ and $ASL = 1/2$.*

Theorem 6.6 (Exact best-case ASL) *When $p = 1$ and t approaches 0, then $\mathcal{A} = g/2$ and $ASL = Ng/2 + 1/2$, where g is the generality.*

Corollary 6.3 (Approximate worst-case ASL) *If $p = 0$ and $t = 1$, then $\mathcal{A} = 1$ and $ASL = N + 1/2$.*

Corollary 6.4 (Exact worst-case ASL) *When $p = 0$ and t approaches 1, then $\mathcal{A} = 1 - g/2$ and $ASL = N(1 - g/2) + 1/2$.*

Example.   A closer examination of Equations 6.5 to 6.7 helps show the relationships and functions of the variables. We assume that $N = 1000$ for this example. Figure 6.2 shows the ASL when retrieval is optimal, that is, when $Q = 1$ and p and t are varied over the full range of possible values. Beginning with the leftmost corner in Figure 6.2 and moving clockwise, we can see that when $p = t = 0$, the $\mathcal{A}$ factor is simply the constant $1/2$, implying essentially random retrieval. Setting p and t to 0 implies that all the documents have

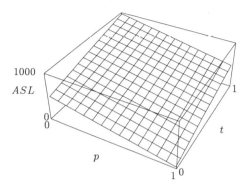

Figure 6.2.   The performance as $p = \Pr(d|rel)$ and $t = \Pr(d)$ vary, decreasing the ASL (increasing the performance) as p increases and t decreases, for $N = 1000$ and $Q = 1$.

feature value 0 and all the relevant documents have value 0; essentially, the relevant documents are not distinguishable from the other documents.

At the top of Figure 6.2, we find that $t = 1$ and $p = 0$, resulting in the weighting of $A = 1/2 - 0/2 + 1/2 = 1$. Almost all the relevant documents have feature frequency 0 while almost all of the remaining documents have feature frequency 1, forcing the relevant documents to the end of the ordering, assuming the ordering $1 \succ 0$.

Continuing clockwise around Figure 6.2, the A when $p = t = 1$ is computed as $A = 1/2 - 1/2 + 1/2 = 1/2$. This is similar to the situation that arises when $p = t = 0$ on the left hand side of the Figure; that is, the relevant documents are distributed in the same class of documents as are all the documents, and the retrieval is essentially random.

The best-case retrieval occurs at the bottom of the Figure when $p = 1$ and $t \to 0$. Here the A component is computed as $1/2 - 1/2 + 0/2 = 0$. There are few documents with the feature but most of the relevant document have the feature, making it easy to isolate the relevant documents.

We compute ASL as $N(QA + \overline{QA}) + 1/2$. One factor in decreasing the ASL is to decrease N. While this is not possible in some circumstances, it is obvious that one will not have to look at as many documents if there are fewer documents to select from, all other factors held equal. Keeping N constant, the ASL is minimized in two cases. The first is when the ranking procedure is optimal and A approaches 0 and the second case is when the ranking procedure is worst-case and A approaches 1.

Variation in ASL values

The variation in ASL with optimal ranking may be used to show how much variation might be expected, given the amount of knowledge available about the parameters. We assume here that t is fully and accurately known. The variation in ASL may be computed by determining the ASL for both very high levels of the prior distribution of p and for low levels of p. For example, if we compute the ASL using the value of p that occurs at the 2.5% point for the prior distribution of p (the point where the CDF for the prior distribution is .025) and also compute the ASL for the 97.5% point, the range between the two different ASLs represents the set of ASL values (or *confidence interval*) over which we have a 95% chance that the actual ASL will occur if we had perfect knowledge.

EXERCISE 6.2. [*Easy*] Assume that we have a database and query such that $N = 10,000,000$, $t = .5$, and that ranking is optimal. What is the ASL when $p = .5$ and what is the ASL at $p = .51$? What is the ASL at $p = .9999$?

EXERCISE 6.3. [*Easy*] Do the preceding exercise assuming that ranking is random and $Q = .5$.

EXERCISE 6.4. [*Moderate*] Assuming that $N = 10,000,000$, $p = .50$, and ranking is optimal, what will be the value of t if the ASL is to be improved so that $ASL = 10$, a reasonable ASL for many human searchers? How do you interpret and explain this result?

EXERCISE 6.5. [*Moderate*] Select 4 terms occurring in this book that you think might be reasonable search terms. Search for these terms on an existing retrieval system or network search engine where you can search for common terms, e.g. "the," to find about the approximate size of the database, or where you know the size of the database through other means. Assume that you are searching for the terms individually and that each query term is in *all* of the relevant documents; it can't get much better than this! Given the empirically determined t values and the assumed $p = 1$, what will be the A and ASL value for each term assuming optimal retrieval. What does this say about the likely search performance?

6.3 A GENERAL THEORY OF PERFORMANCE

A general theory of retrieval performance is developed below consistent with the earlier text filtering and retrieval performance model. The reader who has understood what has been presented so far should find it a logical and comfortable generalization. The general theory allows us to develop an expression relating performance and probabilistic parameters. It is expressed using general probabilities, rather than specific distributions, making it more flexible than what was presented above. We first present the model for document features that are continuous, with an application to normal feature distributions. We then examine a model for discrete feature distributions, including the binary distribution. The Poisson distribution for term frequencies, which has been argued to closely approximate the distribution of terms in natural language, and the negative binomial distribution, which may be similarly used, can be incorporated into the general model.

The expected term frequency for a term in the relevant documents, $E(d|rel) = \mu_r$, is the relative average position (scaled from 0 to 1) for a document in the distribution of relevant documents. By looking at the documents with frequencies at or above this position in all the documents, the $\mathcal{A}$ component may be computed as follows:

$$\mathcal{A} = \int_{\mu_r}^{\infty} \Pr(j)\, dj \qquad (6.8)$$

where $\Pr(j)$ is the probability of a document having feature frequency j.

We use two distributions in developing this model further. We refer to f_{all}, the distribution of feature frequencies in all documents, while f_{rel} refers to the distribution of feature frequencies in relevant documents. The mean of the distribution f_{rel} is μ_r and the mean of the distribution of all documents, f_{all}, is μ, denoted without a subscript.

Definition 6.2 (Survival function) *We denote the sum of the probabilities above frequency x for the distribution of all terms as the* survival function, $S(x, all)$.

The survival function is one minus the cumulative distribution function for the appropriate distribution.

Theorem 6.7 ($\mathcal{A}$ for binary relevance) *For continuous term distributions and binary relevance,*

$$\mathcal{A} = S(\mu_r, all). \qquad (6.9)$$

This follows from the definitions of $\mathcal{A}$ and the survival function.

Continuous Relevance

In some circumstances, it may be beneficial to consider relevance as a continuous variable, with those documents the user wishes to see as the set of documents with associated relevance values at or above a certain point on the relevance scale.

Theorem 6.8 ($\mathcal{A}$ for continuous relevance) *For continuous relevance and relevance scaled from 0 (no relevance) to 1 (complete relevance), then for documents of a relevance value r of x or above, $0 \le x \le 1$, $x \le r \le 1$, with the average frequency for documents of relevance r denoted as μ_r, the average value of $\mathcal{A}$, as computed in Equation 6.9, is*

$$E_x(\mathcal{A}) = \int_x^1 \frac{\Pr(r)}{\int_x^1 \Pr(r')\, dr'}\, S(\mu_r, all)\, dr.$$

The nature of the distribution of μ_r over the set of relevance values is an open and important question.

6.4 CONTINUOUS FEATURE DISTRIBUTIONS

The nature of $\mathcal{A}$ is best understood by examining continuous feature distribution models because of their relative simplicity. We assume binary relevance below unless stated otherwise.

Normal Distribution

The model for $\mathcal{A}$ that is consistent with normally distributed term frequencies may be applied to a variety of feature distributions when used as an approximation. The mean value for the distribution of the features in relevant documents is μ_r, μ is the mean for the distribution of the feature in all documents, and we assume there are equal variances σ^2 for both distributions. We find that

$$S(\mu_r, all) = 1 - \frac{1}{2}\left(1 + Erf\frac{\mu_r - \mu}{\sqrt{2\sigma^2}}\right), \tag{6.10}$$

where $Erf(x)$ is the standard error function,

$$Erf(z) = \frac{2}{\sqrt{\pi}}\int_0^z e^{-t^2}\,dt.$$

For small increments, the increase or decrease in performance is proportional to the factor $(\mu_r - \mu)/\sigma^2$, although the performance does not improve linearly as the factor $(\mu_r - \mu)/\sigma^2$ increases. The reader might note the similarity in structure of this factor and the E measures proposed by Swets (Equation 4.2) and by Brookes (Equation 4.3). This active component in retrieval performance is such that when the difference between μ_r and μ increases, performance improves. As the mean μ_r for the distribution of features in relevant documents increases beyond the mean μ for all documents, presumably most of them nonrelevant, the performance improves because of the increased separation between the feature frequencies for the two categories of documents.

When the variance increases for a fixed set of differences, the performance decreases. This may be best understood by imagining the relevant and all the documents having their means at some fixed difference. If the distributions are narrowed, then there is a less overlap between the two categories of documents and a greater degree of separation. Conversely, if both distributions are nearly flat, that is, both distributions have a high variance, then there is a great deal of overlap between the two distributions, making them harder to separate, resulting in decreased performance.

Similar symbolic and numerical techniques may be used to compute performance values for other continuous distributions, such as the gamma distribution or, more generally, for the exponential family of distributions.

Example. Assume that we are searching using a feature that is normally distributed with $\mu_r = 20$ and $\mu = 10$, with $\sigma^2 = 10$. Using Equation 6.10,

$\mathcal{A} = .159$. If $\mu_r = 15$ and the feature is less discriminating, then we find that $\mathcal{A} = .309$, while if μ_r is further reduced to 10, so that $\mu_r = \mu$, we find that $\mathcal{A} = .5$, which is what one expects from performance that is equivalent to random.

EXERCISE 6.6. [*Moderate*] Develop an equation similar to Equation 6.10 that doesn't assume equal variances. Give an example of when this might prove useful.

6.5 DISCRETE TERM FREQUENCIES

When computing $\mathcal{A}$ using this general approach consistent with discrete term frequencies, a similar discrete technique may be developed with sums rather than integrals. A satisfactory model for the discrete case is similar to the original binary model. This general model will also prove useful later as the basis for multivariate models. $\mathcal{A}$ is computed as

$$\mathcal{A} = 1 - \sum_{i=0}^{\infty} \left(\mathcal{C}_i(D) - \frac{\Pr(d_i)}{2} \right) \Pr(d_i | rel), \qquad (6.11)$$

where

$$\mathcal{C}_n(D) \;=\; \sum_{i=0}^{n} \Pr(d_i). \qquad (6.12)$$

The function $\mathcal{C}_n(D)$ represents the cumulation of probabilities up to the value for n for the random variable D. Using this, Equation 6.11 computes the average position of relevant documents. Note that the majority of Equation 6.11 is subtracted from 1 because it computes an average such that higher values are associated with higher term frequencies, which are, in turn, associated with the lower ASL values.

The continuous analog of Equation 6.11 is used when features d are continuously distributed:

$$A = 1 - \int_{-\infty}^{\infty} \mathcal{C}_i(D) \Pr(d_i | rel) \, di$$

where the $\mathcal{C}_i(X)$ function is the cumulative distribution function $\int_{-\infty}^{i} \Pr(X = x_j) \, dj$.

Binary Model Parameters

Using Equation 6.11, and using the notation $p = \Pr(d = 1|rel)$ and $t = \Pr(d = 1)$, we compute $\mathcal{A}$ for binary terms as

$$\mathcal{A} = 1 - \left[\left((1 - t) - \frac{1 - t}{2}\right)(1 - p) + \left(1 - \frac{t}{2}\right)p\right]$$

$$= \frac{1 - p + t}{2},$$

the result that was computed earlier as Equation 6.1.

Poisson Model Parameters

Estimating the performance of retrieval may assume that term frequencies are distributed as described by the Poisson model. Terms are distributed this way if each occurrence of a term in a document depends only on the instantaneous rate of occurrence of terms, and not on when the previous term occurred. Applying Equation 6.11 to the Poisson distribution results in

$$\mathcal{A} = 1 - \sum_{i=0}^{\infty}\left(\sum_{j=0}^{i}\frac{e^{-\lambda}\lambda^j}{j!} - \frac{e^{-\lambda}\lambda^i}{2(i!)}\right)\frac{e^{-\lambda_r}\lambda_r^i}{i!}. \tag{6.13}$$

Computing the survival function from point x upward for the Poisson distribution may be simplified by using the lower incomplete gamma function, $\gamma(x, n) = \int_0^n e^{-t}t^{x-1}\,dt$. The survival function for the Poisson distribution from x upwards is then $\gamma(x, \lambda)/\Gamma(x)$ (Haight, 1967).

Example. Applying Equation 6.13 to the situation where $\lambda_r = \lambda$ results in $\mathcal{A} = .5$, that is, when the relevant documents are distributed the same as the rest of the documents, then $\mathcal{A}$ reflects the fact that the average relevant document is to be found in the middle of the documents. When $\lambda_r = 4$ and $\lambda = 3$, $\mathcal{A}$ decreases to .355. As the gap between λ_r and λ increases, $\mathcal{A}$ continues to decrease.

EXERCISE 6.7. [*Research Problem*] Assume that term frequencies are truly Poisson distributed and that ranking is optimal. Systems may produce binary valued index terms by assigning a binary 1 when the term is present in the document and a 0 when the term is absent. Compute the decrease in ASL due to binary indexing, when compared to using the full Poisson distributed term frequencies. Note that term frequencies of 0 are the same for both models.

6.6 $\mathcal{A}$, ASL, AND TRADITIONAL PERFORMANCE MEASURES

The ASL measure is computed from the mean location of relevant documents. Other locations could be used, such as the median, the seventy-fifth percentile, or other arbitrary points in the distribution of f_{rel}. One can average the $25\%, 50\%$, and 75% recall points and one can similarly average $\mathcal{A}$ values computed at these points rather than at μ_r.

Given a point x in the distribution of relevant documents, we find that

$$\mathcal{A}(x, all) = \int_x^\infty \Pr(y)\, dy,$$

where Y is the random variable describing term frequencies in all documents. In the case where ASL is to measured, $x = \mu_r$, the mean for the distribution of relevant documents. Other stopping points may represent different stopping rules used by the searcher (Kraft & Lee, 1979). Stopping at μ_r can be interpreted as *satiation* with the relevant documents when reaching this point.

Traditional Performance Measures

The survival function may be used in the definition of traditional performance measures.

Theorem 6.9 (Recall) *The recall at point x, where x is a feature value, may be computed as*

$$recall_x = S(x, rel).$$

This is the percent of the relevant documents retrieved up to point x. If f_{rel} is a symmetrical distribution, $S(\mu_r, rel) = .5$

The precision of a retrieval system at a given recall level x is proportional to the ratio of the number of relevant documents retrieved at that point to the total number of documents retrieved.

Theorem 6.10 (Precision) *The precision at point x is*

$$precision_x = \frac{S(x, rel)}{S(x, all)} g. \tag{6.14}$$

This is the expected number of relevant documents retrieved up to point x divided by the expected number of documents retrieved up to point x. The variable g is the generality measure, $\|\{rel\}\|/\|\{all\}\|$, Equation 4.1. Generality is used here to compensate for unequal numbers of relevant and all documents.

Corollary 6.5 (Precision as a function of recall)

$$precision_x = \frac{recall_x}{S(x, all)} g.$$

This shows the functional relationship between precision and recall.

Note that one can plot a precision-recall graph based on the above formulae to produce a set of (recall, precision) points for each x value, given the generality g and the two distribution functions f_{rel} and f_{all}. One merely moves from $x = 1$

down to $x = 0$, computing the precision and recall values at points sufficiently close together for the level of accuracy desired. One may also numerically solve for one measure from the other by noting that the inverse function of the CDF is the quantile function (found in many computer packages having the CDF), and the inverse of the survival function is one minus the quantile function.

E Measure

Using these measures of precision and recall, one may compute E at point x for optimal ordering thus:

$$
\begin{aligned}
E &= 1 - \frac{2}{\dfrac{1}{P} + \dfrac{1}{R}} \\[2ex]
&= 1 - \frac{2}{\dfrac{1}{g}\dfrac{S(x, all)}{S(x, rel)} + \dfrac{1}{S(x, rel)}} \\[2ex]
&= 1 - \frac{2}{\dfrac{1 + S(x, all)/g}{S(x, rel)}} \\[2ex]
&= 1 - \frac{2S(x, rel)}{1 + S(x, all)/g}.
\end{aligned}
\tag{6.15}
$$

One can compute E as 1 minus the ratio of twice the expected number of relevant documents retrieved up to that point in a search to the sum of the total number of relevant documents and the expected number of documents retrieved up to that point. The effect of generality on the E measure may also be understood using this equation, with an increase in generality decreasing (improving) the E value, with other factors being held constant. Somewhat easier to interpret is the F measure, $F = 1 - E$, which has higher values representing better retrieval results.

Optimal E

The optimal value of x can be computed analytically for some distributions and numerically for most distributions by taking the derivative of Equation 6.15, solving this for 0, and then solving for the other parameters.

One way of studying filtering performance is to examine how the E measure is affected by features distributed in a manner described by the normal distribution. Figure 6.3 illustrates the values of E when the variance of both distributions is set to 3, $\mu = 0$, $x = 4$, and $g = .01$. When the discrimination power is increased, by increasing μ_r (moving to the right on the graph), the E value drops. Interestingly, the performance described by the S shaped curve appears to level off at a point, and there is essentially a floor below which E

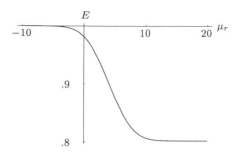

Figure 6.3.   E performance.

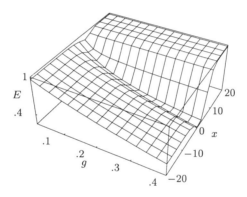

Figure 6.4. E as g and x vary, including the optimal E and its neighboring E values.

does not drop while increasing the discrimination of the term, holding the other parameters constant.

Holding the variance at 3 and setting $\mu_r = 3$ and $\mu = 0$, more complex relationships producing the E value can be studied. Using numerical techniques, for example, for $g = .01$, the optimal E of .9026 occurs when filtering documents with feature frequency of 6.25. Allowing the generality to vary, along with the various document points, one can see in Figure 6.4 how the E values vary as generality and x points vary, and one can see the optimal values for E. Interestingly E seems to drop sharply when x approaches its optimal value from above, but E doesn't increase much as x decreases beyond this point. Generality clearly has a large impact on E, with higher values for generality producing lower E values than are obtained at the optimal x value for lower values of generality.

EXERCISE 6.8. [*Difficult*] What is the expected value of the E measure?

EXERCISE 6.9. [*Research Problem*] What are the variances of the precision and recall measures?

EXERCISE 6.10. [*Research Problem*] What is the variance of the E measure? What is the variance of the optimal E measure?

EXERCISE 6.11. [*Research Problem*] Blair and Maron (1985) note that "the amount of search effort required to obtain the same recall level increases as the database increases, often at a faster rate than the increase in database size." Assuming a single term query, a fixed number of relevant documents, and a fixed p, and for an arbitrary recall level, analytically determine some circumstances under which this does or doesn't hold. Treat ASL as linearly related to "search effort."

6.7   THE EFFECT OF PARAMETER VALUES

When computing the ASL using the methods developed above, each particular set of parameter values will have its own associated ASL value. In our model of text filtering performance, the $\mathcal{A}$ component should always use the exact values that exist in the system, that is, they should reflect perfect knowledge. On the other hand, $\mathcal{Q}$ reflects the quality of the ranking algorithm and incorporates the knowledge held by the user and the ranking algorithm as well as the true values. The parameter $\mathcal{Q}$ captures the relationship between the performance with perfect knowledge, provided by the optimal ranking procedure $\mathcal{R}_o$, and also the sub-optimal knowledge held by the user.

The effect of different parameter values may be seen in Figure 6.5, which shows the ASL for the situation illustrated earlier in Figure 5.2. With parameter q fixed at .3, the ASL varies as parameter p and the weight of knowledge, or confidence in our knowledge about p, is allowed to change. The confidence is computed as the conceptual number of documents retrieved. Parameter $\mathcal{Q}$ is computed as in Equation 5.16. Note that as the confidence about p increases, the ASL improves (decreases).

Ignorance may be reflected in the choice of a sub-optimal ranking algorithm, such as IDF weighting. This is not meant to criticize the use of IDF weighting, which is appropriate in circumstances where the user's lack of knowledge about some parameter values would produce worse performance using other models. Ignorance of parameter values is the second way in which sub-optimal knowledge enters $\mathcal{Q}$.

The Effect of Feedback

The difference between performance with two different levels of relevance feedback may be modeled as $\Delta ASL = ASL_j - ASL_i$, the difference between the performance produced given two different knowledge sets, i and j, representing two different stages in the search, with j assumed to be a later stage than i. The value $\mathcal{A}$ is constant regardless of the human knowledge incorporated into $\mathcal{Q}$.

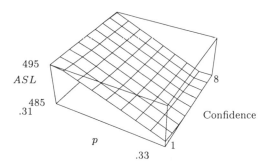

Figure 6.5.   The ASL with $q = .3$ as p and the conceptual strength of our knowledge about p varies.

Theorem 6.11 (Condition for the effectiveness of relevance feedback)
When $A \le 1/2$, relevance feedback that changes the state of knowledge from i to j improves performance as measured by ASL only when $Q_i < Q_j$.

Using either i or j as a subscript to indicate whether the Q is from the i or the j state of nature, we find that

$$\Delta ASL = ASL_j - ASL_i$$
$$= N \left[Q_j A + \overline{Q}_j \overline{A} - Q_i A - \overline{Q}_i \overline{A} \right]$$
$$= (Q_j - Q_i)A + (\overline{Q}_j - \overline{Q}_i)\overline{A}$$
$$= (Q_j - Q_i)2(A - \frac{1}{2}).$$

Feedback will improve performance (decrease ASL) when ΔASL is negative, that is, the ASL changes by the amount ΔASL. Performance will always increase when

1. $Q_j > Q_i$ when $A < 1/2$, or

2. $Q_j < Q_i$ when $A > 1/2$.

Clearly, the first condition is what is found when the term is a positive discriminator and feedback increases Q.

EXERCISE 6.12. [*Research Problem*] The rate of ASL increase as Q increases is $N(2A - 1)$. Thus, for $A < 1/2$, ASL decreases as Q increases. For a binary model, where knowledge about p for relevant documents is beta distributed, what is the rate of increase in the ASL as the sample size increases?

6.8 SUMMARY

Text filtering and retrieval performance may be computed after documents are retrieved, based on the results obtained from actual searches, or performance may be predicted beforehand based on probabilistic considerations. Simulations in the case of single term queries are unnecessary; the expected results may be computed analytically using simple methods. To use these techniques, more research is needed on the exact form of probabilistic distributions describing term frequencies in documents. Inconsistencies in experimental results, where the best method with one database is not the best with another, is due in part to the differences in these distributions. The design of a system should incorporate what is known about the distribution of term frequencies when designing the ranking algorithm and when predicting the performance.

In the following chapters, we examine the multivariate techniques necessary to understand multi-term queries and more complicated relationships, including the role of natural language processing in text retrieval and filtering.

7 MULTIVARIATE PROBABILITIES

The probability that one party attaches to cooperative behavior by other parties
—Definition of "trust," Hwang and Burgers, 1997

The probabilistic models of text filtering and retrieval discussed so far have been limited to a single term for the query or filtering specification. While this has allowed us to examine some basic relationships and phenomena in filtering, extensions of probabilistic models are needed to understand multi-term retrieval models. The inter-term relationships are often studied through linguistic or knowledge-based analyses, and an understanding of the effect of these techniques requires the knowledge of suitable multivariate techniques. Below, we consider useful methods for estimating probabilities for more than a single term at a time. The problem is thus how to estimate $\Pr(\boldsymbol{d})$, where $\boldsymbol{d}$ is a vector of n different terms, $\boldsymbol{d} = \{d_1, d_2, \dots, d_n\}$. This chapter considers how to estimate $\Pr(\boldsymbol{d})$ without assuming term independence.

There are few simple and elegant models of multivariate relationships for cases other than the multivariate normal distribution. Multivariate binary and Poisson distributions are not as efficiently brought together into multivariate models incorporating statistical dependence between terms as are the multivariate normal models (Johnson, Kotz, & Balakrishnan, 1997). Many of these discrete distributions have no simple conjugate prior distributions supporting both the term distributions and the associations between terms. Several multivariate binary models will be examined below, although with little discussion about prior information. While numerical techniques are available and are used below to show performance, the multivariate normal distribution is the model which is most fully developed in a Bayesian manner. There is a paucity of multivariate normally distributed data in retrieval systems; its primary use here is to help explain the performance model and to be used as an approximation for multivariate Poisson and negative binomial distributions.

A major practical problem with using multivariate statistical models in text filtering and retrieval systems with large numbers of terms is the difficulty in accurately estimating the probabilities of specific combinations of terms that do not occur in the database used in developing the initial probabilities. Accurately estimating joint probabilities consistent with multivariate assumptions is almost impossible for relevant documents, because of the very small number of documents for which relevance feedback judgements are available and from which one can learn. Other sources for accurate estimates are linguistic and knowledge-based relationships.

The inter-term relationship most frequently described below is the *covariance* between terms. Informally, covariance is how one term varies as another term varies. This is often normalized to the Pearson product moment *correlation*, a value that is within the range of -1 to $+1$.

7.1 MULTIVARIATE BINARY DISTRIBUTIONS

In most text filtering and retrieval systems, term occurrences are treated as binary events, either occurring or not occurring. These occurrences, and the relationships between them, may be found using a multivariate model proposed by Teugel (1990).

Terms are numbered here from right to left, so that the rightmost of n terms is numbered 1 and the leftmost is term n. The joint probabilities of term 1 having value i and term 2 having value j is denoted as p_{ji}. The probabilities of the four possible combinations of the two terms may be computed as:

$$p_{00} = (1 - p_1)(1 - p_2) + \sigma_{12} = 1 - p_1 - p_2 + \mu_{12}$$
$$p_{01} = p_1(1 - p_2) - \sigma_{12} = p_1 - \mu_{12}$$
$$p_{10} = (1 - p_1)p_2 - \sigma_{12} = p_2 - \mu_{12}$$
$$p_{11} = p_1 p_2 + \sigma_{12} = \mu_{12}$$

Two values are used here to study the association between the two terms. The first is the covariance (σ) between the two, that is, how they vary with each other. Computed as

$$\sigma_{12} = E\left[(D_1 - p_1)(D_2 - p_2)\right],$$

it is the expected value of the difference between each value and its probability, or between each value and the mean (which is p for a binary distribution). The covariance is widely used in other areas of statistics and its use here provides a link between the analytic method and other statistical techniques. The co-variance between n terms, numbered $i, j, \ldots, z$, and with means $\mu_i, \mu_j, \ldots, \mu_z$, will be denoted as

$$\sigma_{ij\ldots z} = E\left[(D_i - \mu_i)(D_j - \mu_j) \cdots (D_z - \mu_z)\right].$$

The covariance may be transformed into the more familiar correlation mea-sure in the case of binary data through use of the following:

$$\rho_{12} = \frac{\sigma_{12}}{\sqrt{p_1(1 - p_1)p_2(1 - p_2)}}, \tag{7.1}$$

with an obvious generalization to more than two cases. The denominator in this conversion from covariance to correlation is the variance of the two terms and acts as a normalizing factor, limiting the correlation coefficient to the range from -1 to $+1$ for all distributions.

A correlation of 0 between a pair of terms indicates that the variation in the frequency of one term provides no information about the variation in the frequency of the other term. A perfect correlation of 1 occurs when the variation in one term provides complete information about the variation in the frequency of another. For example, if two coins were glued together at their edges, side by side, so that both faces showed at the same time, flipping them and knowing that one was a *heads* would tell us with certainty that the second coin was also a *heads*. Conversely, the correlation of -1 tells us that we know with certainty that when one coin is *heads* the other will be *tails*, such as would occur if the coins were glued together side by side so that when *heads* is seen for one coin, *tails* is seen for the other.

Using Equation 7.1, the range of possible covariance values for two terms may be computed. Setting the correlation ρ to 1,

$$\sigma_{max} = \sqrt{p_1(1 - p_1)p_2(1 - p_2)}.$$

Setting the correlation ρ to -1 produces

$$\sigma_{min} = -\sqrt{p_1(1 - p_1)p_2(1 - p_2)}.$$

This may be easily generalized to n-way correlations and covariances.

EXERCISE 7.1. [*Easy*] Assume 12 documents, with 5 documents having both the terms *text* and *filtering*. Three documents have just *text*, 2 just *filtering*, and 2 doc-uments have neither term. What is the covariance between the two terms? What is the correlation?

7.2   A MATRIX MODEL OF MULTIVARIATE BINARY DATA

The use of matrices in describing binary term dependencies will simplify the presentation, as well as allowing for its generalization over a large number of terms. In addition to using traditional matrix multiplication, we will also use the outer, or Kronecker, product, denoted as $\otimes$. A matrix m which is being multiplied by itself n times using the outer product is denoted as $\mathbf{m}^{\otimes n}$. An array holding the information for n terms is often denoted below with a superscript "(n)".

Kronecker Product

The Kronecker or outer product may be seen as a useful form of multiplication for arrays. Assume we begin with two matrices,

$$A = \begin{bmatrix} a_{11} & a_{12} \\ a_{21} & a_{22} \end{bmatrix} \qquad B = \begin{bmatrix} b_{11} & b_{12} \\ b_{21} & b_{22} \end{bmatrix}.$$

The outer product of these is

$$A \otimes B = \begin{bmatrix} a_{11}b_{11} & a_{11}b_{12} & a_{12}b_{11} & a_{12}b_{12} \\ a_{11}b_{21} & a_{11}b_{22} & a_{12}b_{21} & a_{12}b_{22} \\ a_{21}b_{11} & a_{21}b_{12} & a_{22}b_{11} & a_{22}b_{12} \\ a_{21}b_{21} & a_{21}b_{22} & a_{22}b_{21} & a_{22}b_{22} \end{bmatrix}.$$

This differs from "standard" matrix multiplication, where

$$AB = \begin{bmatrix} a_{11}b_{11} + a_{12}b_{21} & a_{11}b_{12} + a_{12}b_{22} \\ a_{21}b_{11} + a_{22}b_{21} & a_{21}b_{12} + a_{22}b_{22} \end{bmatrix}.$$

Probabilities with Two Terms

The set of probabilities for the values p_{00} through p_{11} are represented by $\mathbf{p}^{(2)}$, with the probabilities being stored in this array in increasing order of the subscripts, (i.e., 00, 01, 10, 11). Similarly, the covariance matrix σ is used to store all the covariance values for the same subscripts as with the $\mathbf{p}$ array. The array $\sigma = (1, 0, 0, \sigma_{12})^T$ provides the covariance for no terms, then for the rightmost term alone, then for the left term alone, and then for both terms.

The set of probabilities for different document profiles may be computed as

$$\mathbf{p}^{(2)} = \begin{bmatrix} q_2 & -1 \\ p_2 & 1 \end{bmatrix} \otimes \begin{bmatrix} q_1 & -1 \\ p_1 & 1 \end{bmatrix} \begin{bmatrix} 1 \\ 0 \\ 0 \\ \sigma_{12} \end{bmatrix}.$$

If we choose to use the vector of means μ for each set of subscripts, we may compute $\mu^{(2)} = (1, p_1, p_2, \mu_{12})^T$ which represents the mean values for no features, feature 1 only, feature 2 only, and both features together. Using this, we may compute $\boldsymbol{p}$ as

$$\boldsymbol{p}^{(2)} = \begin{bmatrix} 1 & -1 \\ 0 & 1 \end{bmatrix} \otimes \begin{bmatrix} 1 & -1 \\ 0 & 1 \end{bmatrix} \begin{bmatrix} 1 \\ p_1 \\ p_2 \\ \mu_{12} \end{bmatrix}.$$

Consider the more general case where the method is generalized to n terms. Both of the above methods of computing the probability vector $\boldsymbol{p}$ may be generalized to n terms. One may compute $\boldsymbol{p}^{(n)}$ as

$$\boldsymbol{p}^{(n)} = \begin{bmatrix} 1 & -1 \\ 0 & 1 \end{bmatrix}^{\otimes n} \boldsymbol{\mu}^{(n)}.$$

Similarly, one may compute the mean array $\boldsymbol{\mu}^{(n)}$ as

$$\boldsymbol{\mu}^{(n)} = \begin{bmatrix} 1 & 1 \\ 0 & 1 \end{bmatrix}^{\otimes n} \boldsymbol{p}^{(n)}, \tag{7.2}$$

where the superscript in $\boldsymbol{x}^{\otimes n}$ represents the array $\boldsymbol{x}$ taken to the nth power, with the multiplication using the outer product.

Using the covariance array $\boldsymbol{\sigma}^{(n)}$, one may compute $\boldsymbol{p}^{(n)}$, in the case of n terms, as

$$\boldsymbol{p}^{(n)} = \begin{bmatrix} q_n & -1 \\ p_n & 1 \end{bmatrix} \otimes \begin{bmatrix} q_{n-1} & -1 \\ p_{n-1} & 1 \end{bmatrix} \otimes \cdots \otimes \begin{bmatrix} q_1 & -1 \\ p_1 & 1 \end{bmatrix} \boldsymbol{\sigma}^{(n)}.$$

The covariance array may itself be computed from $\boldsymbol{p}^{(n)}$

$$\boldsymbol{\sigma}^{(n)} = \begin{bmatrix} 1 & 1 \\ -p_n & q_n \end{bmatrix} \otimes \begin{bmatrix} 1 & 1 \\ -p_{n-1} & q_{n-1} \end{bmatrix} \otimes \cdots \otimes \begin{bmatrix} 1 & 1 \\ -p_1 & q_1 \end{bmatrix} \boldsymbol{p}^{(n)}. \tag{7.3}$$

Example. Consider the data in Table 7.1, where the author excluded profiles with a "1 0" for the right two terms and doubled the probabilities for those documents with a "1 0" for the left two terms.

Table 7.1.   Set of 8 document profiles, term frequencies and related parameters.

Profile	Frequency	$\Pr(d_i)$	σ	μ
0 0 0	1	1/8	1	1
0 0 1	1	1/8	0	5/8
0 1 0	0	0	0	1/4
0 1 1	1	1/8	3/32	1/4
1 0 0	2	1/4	0	5/8
1 0 1	2	1/4	-1/64	3/8
1 1 0	0	0	-1/32	1/8
1 1 1	1	1/8	-1/128	1/8

The vector μ may be computed through the application of Equation 7.2 to the data in Table 7.1.

$$\mu^{(3)} = \begin{bmatrix} 1 & -1 \\ 0 & 1 \end{bmatrix}^{\otimes 3} p^{(3)}$$

$$\begin{bmatrix} 1 \\ 5/8 \\ 1/4 \\ 1/4 \\ 5/8 \\ 3/8 \\ 1/8 \\ 1/8 \end{bmatrix} = \begin{bmatrix} 1 & 1 & 1 & 1 & 1 & 1 & 1 & 1 \\ 0 & 1 & 0 & 1 & 0 & 1 & 0 & 1 \\ 0 & 0 & 1 & 1 & 0 & 0 & 1 & 1 \\ 0 & 0 & 0 & 1 & 0 & 0 & 0 & 1 \\ 0 & 0 & 0 & 0 & 1 & 1 & 1 & 1 \\ 0 & 0 & 0 & 0 & 0 & 1 & 0 & 1 \\ 0 & 0 & 0 & 0 & 0 & 0 & 1 & 1 \\ 0 & 0 & 0 & 0 & 0 & 0 & 0 & 1 \end{bmatrix} \begin{bmatrix} 1/8 \\ 1/8 \\ 0 \\ 1/8 \\ 1/4 \\ 1/4 \\ 0 \\ 1/8 \end{bmatrix}.$$

As a check of this, note that the second position (μ_{001}) is 5/8, the correct value for p_1. Similarly, $\mu_{010} = 1/4 = p_2$ and $\mu_{100} = 5/8 = p_3$.

The covariance vector can be similarly computed using Equation 7.3 as

$$\sigma^{(n)} = \begin{bmatrix} 1 & 1 \\ -\frac{5}{8} & \frac{3}{8} \end{bmatrix} \otimes \begin{bmatrix} 1 & 1 \\ -\frac{1}{4} & \frac{3}{4} \end{bmatrix} \otimes \begin{bmatrix} 1 & 1 \\ -\frac{5}{8} & \frac{3}{8} \end{bmatrix} p^{(n)}$$

$$\begin{bmatrix} 1 \\ 0 \\ 0 \\ \frac{3}{32} \\ 0 \\ -\frac{1}{64} \\ -\frac{1}{32} \\ -\frac{1}{128} \end{bmatrix} = \begin{bmatrix} 1 & 1 & 1 & 1 & 1 & 1 & 1 & 1 \\ -\frac{5}{8} & \frac{3}{8} & -\frac{5}{8} & \frac{3}{8} & -\frac{5}{8} & \frac{3}{8} & -\frac{5}{8} & \frac{3}{8} \\ -\frac{1}{4} & \frac{3}{4} & -\frac{1}{4} & \frac{3}{4} & -\frac{1}{4} & \frac{3}{4} & -\frac{1}{4} & \frac{3}{4} \\ \frac{5}{32} & -\frac{3}{32} & -\frac{15}{32} & \frac{3}{32} & \frac{5}{32} & -\frac{3}{32} & -\frac{15}{32} & \frac{3}{32} \\ -\frac{5}{8} & -\frac{5}{8} & -\frac{5}{8} & -\frac{5}{8} & \frac{3}{8} & \frac{3}{8} & \frac{3}{8} & \frac{3}{8} \\ \frac{25}{64} & -\frac{15}{64} & \frac{25}{64} & -\frac{15}{64} & -\frac{15}{64} & \frac{9}{64} & -\frac{15}{64} & \frac{9}{64} \\ \frac{5}{32} & \frac{15}{32} & -\frac{5}{32} & -\frac{15}{32} & -\frac{3}{32} & -\frac{9}{32} & \frac{3}{32} & \frac{9}{32} \\ -\frac{25}{256} & \frac{15}{256} & \frac{75}{256} & -\frac{45}{256} & \frac{15}{256} & -\frac{9}{256} & -\frac{45}{256} & \frac{27}{256} \end{bmatrix} \begin{bmatrix} 1/8 \\ 1/8 \\ 0 \\ 1/8 \\ 1/4 \\ 1/4 \\ 0 \\ 1/8 \end{bmatrix}.$$

Using the covariance matrix $\sigma^{(3)}$ one may compute the correlation vector as

$$\rho = (1, 0, 0, 0.447, 0, -0.066, -0.149, -0.077)^T,$$

showing the relationships between terms. The strongest correlation, $\rho_{12} = .447$, is between terms 1 and 2, the rightmost terms. This is due primarily to the way in which the author forced there to be no occurrences of documents with terms 1 and 2 being 10. The correlation between all three terms, $\rho_{123} = -.077$, is in addition to the pairwise correlations and captures the interaction between the three terms that goes beyond the pairwise relationships.

EXERCISE 7.2. [*Moderate*] Using the data in the previous exercise (page 131), compute $\mu^{(2)}$ and $\sigma^{(2)}$.

7.3   A BINARY MULTIVARIATE EXPANSION

The Bahadur-Lazarsfeld Expansion (BLE) may be used to estimate probabilities given binary features and feature frequencies (Bahadur, 1961). This is an alternative to the matrix method proposed by Teugel above. Controlling the degree of expansion of the BLE between degrees 1 and n allows one to incorporate varying degrees of dependence. Here, expansion to level 1 produces probabilities consistent with data independence, while using the expansion to level n, where n is the number of document features, provides full dependence information. The probability of a particular profile d is

$$\Pr(d) = \prod_{i=1}^{t} p_i^{d_i}(1-p_i)^{(1-d_i)}\Big[1+$$

$$\sum_{i<j} \rho_{ij} \frac{(d_i - p_i)(d_j - p_j)}{\sqrt{p_i p_j(1-p_i)(1-p_j)}} +$$

$$\sum_{i<j<k} \rho_{ijk} \frac{(d_i - p_i)(d_j - p_j)(d_k - p_k)}{\sqrt{p_i p_j p_k(1-p_i)(1-p_j)(1-p_k)}} +$$

$$\rho_{12...t} \frac{(d_1 - p_1)(d_2 - p_2)\ldots(d_t - p_t)}{\sqrt{p_1 p_2 \ldots p_t(1-p_1)(1-p_2)\ldots(1-p_t)}}\Big]. \quad (7.4)$$

The BLE may be truncated at any point, limiting the degree of dependence (and accuracy) that is incorporated into estimating the probabilities. If we consider the data from Table 7.1 (page 134), the probabilities may be estimated using varying degrees of dependence:

$\Pr(d_i, d_j, d_k)$	3-Way Probability	2-Way Probability	Independent Probability
$\Pr(0,0,0)$	1/8	15/128	27/256
$\Pr(0,0,1)$	1/8	17/128	45/256
$\Pr(0,1,0)$	0	1/128	9/256
$\Pr(0,1,1)$	1/8	15/128	15/256
$\Pr(1,0,0)$	1/4	33/128	45/256
$\Pr(1,0,1)$	1/4	31/128	75/256
$\Pr(1,1,0)$	0	-1/128	15/256
$\Pr(1,1,1)$	1/8	17/128	25/256
Sum:	1	1	1

The rightmost column here illustrates the probability that would be computed assuming statistical independence of features. These independent probabilities differ from those using three-way dependence. The three-way dependencies are those exact probabilities found in Table 7.1 and should be considered the "correct" values; the degree to which estimates approach these correct values serves as an indicator of the quality of the estimates.

If we consider the probability of a document having the profile 010, which is 0 under the assumption of full dependence and full knowledge, the probability is estimated using the BLE and term independence as 9/256. When the assumption of two-way dependence is made, the probability is more accurately estimated as $1/128 (= 2/256)$. The change from 9/256 to 2/256 to 0/256 shows a consistent movement toward the correct value, showing increasing accuracy.

Besides the general problem of inaccurate estimates, another drawback of truncating the BLE at an arbitrarily chosen degree of dependence is the possibility of producing impossible probabilities that are outside the range of 0 to 1. Consider the profile 110, which has a probability of 0. Assuming the statistical independence of features produces an estimate of the probability of 15/256. Assuming pairwise dependence results in an estimate of the probability of $-1/128$. Negative probabilities such as this are not meaningful or valid probabilities; similarly, probabilities of 1 or over are invalid. It is clear that the BLE can generate invalid probabilities, such as $-1/128$. The possibility that the BLE can produce such improper probabilistic estimates is a major problem with the use of this expansion.

7.4   TREE-BASED DEPENDENCE

One solution to the limitations imposed by the BLE is the use of a tree-based dependence model. Such an algorithm may begin with a tree structure, where each term on the tree is understood to be linked to those neighboring terms on the tree that have the greatest degree of mutual information between the terms (Chow & Liu, 1968; Van Rijsbergen, 1977). Limiting the dependencies to the more important relationships may decrease the processing time for a real-time system, as well as limiting the estimation errors introduced into the probability to those related to important probabilities, removing the errors associated with

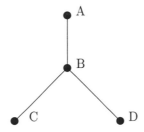

Figure 7.1.   Maximum spanning tree.

less important relationships. While using the probabilities from only these related terms may limit the amount of dependence information available, when
compared to the dependence information used with the BLE, the probabilities
used from such trees are informally understood as the most important relationships and thus the most important term dependencies. This should result
in the most important dependencies being used, maximizing the dependence
information incorporated for the time spent incorporating dependencies and
maximizing the benefit available from this potentially time consuming activity.
Yu et al (1983) propose a more generalized tree structure that can be used for
n-way dependencies that may combine the best of tree and BLE based methods. At present, networks, including Bayesian belief networks, are probably
the most commonly used means of including partial dependencies.

A tree may be constructed so that the mutual information between nodes
on the tree and neighboring nodes is maximized, as in Figure 7.1 (Chow & Liu,
1968). Such a Maximum Spanning Tree (MST) maximizes

$$\sum_{i,j} I(Node_i, Node_j),$$

where $I(i,j)$ represents the expected mutual information provided by node i
about node j,

$$I(i,j) = \sum_{i,j} \Pr(i,j) \log \frac{\Pr(i,j)}{\Pr(i)\Pr(j)}.$$

The probability for each node in the tree may be estimated in isolation, or
it may be estimated conditioned on the node above it on the tree. Given
the four features A through D shown in 7.1, we can estimate probabilities for
terms such as B as either $\Pr(B)$, which assumes independence of terms, or
$\Pr(B = b | A = a)$ which incorporated the dependence information provided by
A about B. If we have a document with two features, B and C, the probability
may be computed as

$$Pr(B,C) = \Pr(B = 1 | A = 0) \Pr(C = 1 | B = 1) \Pr(D = 0 | B = 1) \Pr(A = 0).$$

Prior Information

The prior distribution for a set of independent binary values is the Dirichlet distribution, which can be used as a multivariate generalization of the beta distribution (Gelman, Carlin, Stern, & Rubin, 1995; Johnson et al., 1997). Dependence information isn't modeled by the Dirichlet prior and there is no conjugate prior distribution that can fully capture prior information about multivariate binary dependent data. One is limited in this situation to using an empirical, non-conjugate prior distribution if one wishes to model feature dependence, or one can use a conjugate prior and explicitly assume feature independence.

EXERCISE 7.3. [*Moderate*] Using the data in the first exercise in this chapter (page 131), compute the probability for each of the 4 possible outcomes using the Bahadur Lazarsfeld Expansion.

7.5 BAYESIAN NETWORKS

Relationships between causal variables are frequently represented by a Bayesian belief network. Relationships in these graphs are used in estimating probabilities. Information that A (e.g., representing whether the sky is overcast or not) causes (fully, or in part) B, which perhaps represents whether it is raining or not, and B causes C, the latter perhaps representing whether plants are growing at a high rate of speed. Each node on such a network is caused only by those nodes immediately above it (there may be more than one). In this case A causes B, and A only indirectly causes C; for the purposes of a belief network, all the influence A has on C is through B, allowing us to say that C is caused only by B.

Analysis of a Bayesian network might begin by learning of a value being assigned to a network node. For example, it might be reported that it is overcast outside. Given this and no other observations about whether it is raining or whether plants are growing at a high rate of speed, the added knowledge about overcast conditions can be used to modify our estimate of the probability that it is raining. This probability is influenced by three factors; the probability or knowledge concerning whether it is overcast, the prior probability that it is raining, and the probability or knowledge concerning whether plants are growing at a high rate of speed.

Two forms of influence or probabilities are found here. One, which we refer to as a *forward* influence, is provided about the probability it is raining by the information about whether it is overcast. *Backward* influence is provided by our knowledge concerning whether plants are growing at a high rate of speed which affects our estimate of the probability that it is raining, a causal factor. This form of backward influence is used when one sees someone entering a building with a wet raincoat. Given a commonly accepted causal relationship between rain falling from the sky and wet spots appearing on the clothing of unsheltered people out-of-doors, one revises one's estimate of the probability that it is raining outside. One is working from a result back to the cause.

A Bayesian belief network may be used to estimate probabilities that terms occur in relevant documents, based on query terms or relevance feedback (Turtle & Croft, 1991; Callan, Lu, & Croft, 1995; Ribeiro & Muntz, 1996). First, the network itself is constructed, given data about term relationships derived from a database. Next, for each query or possibly for each document judged relevant, those terms in this query or relevant document are instantiated to 1 in the network. Several passes are made through the network, propagating information in both directions through the belief network. Each document to be considered for retrieval is then evaluated, based on the probabilities in the network, each probability being used as the basis for weights for each document term. Documents may then be assigned a document weight and the documents ranked.

Prior distributions can be used to represent knowledge about the state of individual nodes. Individual binary probabilities may use the beta distribution as the prior distribution, as was found with the univariate distribution. In cases where one wishes to treat learning from multiple samples about multiple nodes, the Dirichlet prior can be used (Heckerman, Geiger, & Chickering, 1995). The problem of fully capturing dependencies in this environment remains to be solved, but this is not a major problem in many Bayesian network applications, because the parental nodes can be treated as though they are statistically independent of each other.

7.6 MUTUAL INFORMATION

The information that one term contains about another term may be measured using the *expected mutual information measure* (EMIM). This may be applied to any two terms (d_i and d_j) or, more generally, probabilistic distributions. The self-information about a particular event occurring with probability p is $-\log p$ (Losee, 1997c). When the logarithm is taken to base 2, the information is measured in bits. The information that text characteristics d_i and d_j provide about each other is

$$I(d_i, d_j) = \log_2 \frac{\Pr(d_i, d_j)}{\Pr(d_i)Pr(d_j)}.$$

Using this, the EMIM may be calculated as

$$\text{EMIM}(d_i, d_j) = \sum \Pr(d_i, d_j) \log_2 \frac{\Pr(d_i, d_j)}{\Pr(d_i)Pr(d_j)}. \tag{7.5}$$

This represents the average information provided by features d_i and d_j about each other. The value $I(d_i, d_j)$ addresses only the information about a single set of values, e.g., $d_i = 1$, $d_j = 0$. EMIM, on the other hand, computes the average $I()$ over the set of all possible values that d_i and d_j may take.

In terms of the 2×2 tables discussed in Chapter 3, the EMIM may be expressed as

$$\mathrm{EMIM}(d_i, d_j) = \log_2 \frac{a(a + b + c + d)}{(a + b)(a + c)}.$$

This can be compared to the ϕ^2 measure of association (Conrad & Utt, 1994) often used with 2×2 tables, which may be computed as

$$\phi^2 = \frac{(ad - bc)^2}{(a + b)(a + c)(b + d)(c + d)}.$$

EXERCISE 7.4. [*Easy*] Using the data in the first exercise in this chapter (page 131), compute the expected mutual information between *text* and *filtering*.

7.7   LOGISTIC MODELS

We have seen the use of the logarithm of odds when developing the binary independence retrieval model. The log-odds is sometimes referred to as the logit transformation of $\Pr(x)$, or

$$\mathrm{logit}(X) = \ln \frac{\Pr(X = 1)}{1 - \Pr(X = 1)} = \alpha + \beta_1 x_1 + \beta_2 x_2 + \cdots + \beta_n x_n. \tag{7.6}$$

This may be understood in relation to the odds of X and was found earlier when considering the odds that a document has a particular value (e.g., Equation 3.9.) In statistical packages, the α and β components in Equation 7.6 are estimated and this equation is itself an approximation, as are many of the equations in this section. Techniques for estimating parameters α and β are described in many statistics books and are implemented in most statistical packages.

One can move from probabilities to the logit value and then to odds: they are different ways of addressing the same problem:

$$\Pr(X = 1) = \frac{O(X = 1)}{1 + O(X = 1)}.$$

Using the odds, the logit transformation can be introduced into the logistic model as

$$O(X = 1) = e^{\mathrm{logit}(X)} = e^{\ln(O(X=1))}.$$

The exponent X has the two components found earlier, a constant, usually written as α, and $\sum_i \beta_i x_i$.

The standard logistic distribution (density function) is

$$Pr(x) = \frac{e^{-x}}{(1 + e^{-x})^2}.$$

The logistic distribution has the frequently found sigmoid shaped cumulative distribution function

$$C_x(X) = \frac{1}{(1 + e^{-x})}.$$

This distribution is similar to that obtained with the normal distribution, except at the tails where it varies significantly (Press, 1972). The cumulative logistic function $f(x)$ can take on a set of values from 0 to 1:

$$C_{-\infty}(X) = \frac{1}{1 + e^{-(-\infty)}} = 0 \qquad C_{+\infty}(X) = \frac{1}{1 + e^{-(+\infty)}} = 1$$

The survival function (and thus $\mathcal{A}$) is

$$\mathcal{A}_x = \mathcal{S}_x(X) = 1 - \frac{1}{1 + e^{-x}}. \tag{7.7}$$

Dependent values may enter logistic models through the introduction of additional variables. These *interaction* variables are included along with the individual variables, allowing the indirect introduction of dependence. For example, one might include the individual terms *text* or *filtering*, as well as the phrase *text filtering* treated as a single unit. The logistic model can capture the same kinds of dependencies that are obtained with the matrix model or the BLE that were discussed earlier, but the logistic model may be computationally faster in many circumstances.

The biggest advantage to using the logistic model for binary data in analytic models of filtering is the presence of an analytically tractable survival function. This enables us to easily model the $\mathcal{A}$ value for binary data with dependencies. Conversely, while this model is simple, it is inexact and makes some simplifying assumptions that often are not met by the data or the system. It can be used for numerical analysis of $\mathcal{A}$.

EXERCISE 7.5. [*Moderate*] Use a statistical package and the data from the first exercise in the chapter to determine the α and β coefficients. Using this, compute $\mathcal{A}$ for this data.

7.8 MULTIVARIATE NORMAL DISTRIBUTION

The most straightforward and tractable multivariate distribution is the *multivariate normal distribution*, also referred to as the *multinormal distribution*. The special cases where there are two or three variables are referred to as the *bivariate* and *trivariate* normal distributions (Cox & Wermuth, 1991). There are several fast numerical approximations available for the calculation of the density and distribution functions for the two- and three-term case, as well as the more general n term case (Lee, 1989, p. 175).

Bayesian analysis of the multivariate normal is simpler than for other distributions because of the existence of conjugate priors. As with the univariate normal distribution, the prior knowledge of the means for each distribution is normally distributed. The inverse-gamma distribution is the conjugate prior for the univariate variance, while the inverse-Wishart distribution is the conjugate prior distribution for the covariance matrix (Johnson & Kotz, 1972; Gelman et al., 1995). The Wishart distribution is a multivariate generalization of the gamma distribution, and using the inverse Wishart as the conjugate prior may be viewed as a multivariate generalization of the use of the inverse gamma distribution as the conjugate prior distribution for the normal distribution's variance.

Our consideration of the general case for the multivariate normal distribution, where there are n terms, may begin with the case where terms are assumed to be statistically independent. The probability of obtaining x when the random variables are *unit normal variables* is (Johnson & Kotz, 1972)

$$\Pr(x) = \frac{1}{(2\pi)^{n/2}} e^{-\frac{1}{2}x^T I x}, \tag{7.8}$$

where n is the number of items in term vector x and I is the identity matrix, essentially a covariance matrix with all covariances 0 except for the covariance of a variable with itself, which is 1.

If we introduce term dependence, represented by a covariance vector σ, and do not assume normalized variables, the probability of a particular set of n values x is calculated as

$$\Pr(x) = \frac{1}{(2\pi)^{n/2}|\sigma|^{\frac{1}{2}}} e^{-\frac{1}{2}(x-\mu)^T \sigma^{-1}(x-\mu)}, \tag{7.9}$$

where μ is the is the mean vector, σ is the covariance matrix, and $|\sigma|$ is the determinant of the covariance matrix.

Multivariate Normal Survival Function

The survival function for the multivariate normal distribution (Drezner, 1992), used to calculate $\mathcal{A}$, is computed as

$$\mathcal{S}_x(X) = \frac{1}{(2\pi)^{n/2}|\sigma|^{1/2}} \int_{x_n}^{+\infty} \cdots \int_{x_1}^{+\infty} e^{-\frac{1}{2}(X-\mu)^T \sigma^{-1}(X-\mu)} \, dX_1 \cdots dX_n \tag{7.10}$$

where n is the number of distributions and x represents the vector of cutoff values.

This survival distribution is not very simple and it may be most easily used through the application of numerical analysis rather than through finding symbolic solutions illustrating performance.

A Prior Distribution

The multivariate normal distribution has two types of parameters: those representing the mean and those describing the variance. The conjugate prior for the means is represented by a normal distribution and the conjugate prior for the variance values is the inverse-chi-square distribution for the univariate case and the inverse-Wishart distribution for the multivariate estimation of the variance. The joint distribution of the normal and the inverse-Wishart distribution is the conjugate distribution for the the multivariate normal distribution (Gelman et al., 1995, p. 80).

The values for the posterior parameters, given the prior parameters, are

$$\mu_n = \frac{n_i}{n_i + n}\mu_i + \frac{n}{n_i + n}\bar{x}$$

$$n_n = n_i + n$$

$$\nu_n = \nu_i + n$$

$$\Lambda_n = \Lambda_i + \sum_{i=1}^{n}(x_i - \bar{x})(x_i - \bar{x})^T + \frac{n_i n}{n_i + n}(\bar{x} - \mu_i)(\bar{x} - \mu_i)^T.$$

Parameter μ_i represents the prior mean. The Λ components represent the inverse of the covariance matrix, referred to as the precision matrix: Λ_i is the initial precision matrix, with Λ_n being the precision matrix after n sample items have been examined. Parameter n_i is the number of prior measurements, n the number of items in the sample. The mean of the sample is denoted as $\bar{x}$, with each of the n data items denoted as $x_1, x_2, \ldots, x_n$. The ν values represent "degree of freedom" components.

Multivariate Normal Approximations of Term Frequencies

Multivariate models of term frequencies that assume terms are Poisson or negative binomially distributed usually assume independence of term frequencies. This defeats one purpose of multivariate models, which are most accurate when they incorporate the relationships between terms. Term independence and the term frequency models may be used in such a way that the multivariate model is simply the univariate model repeated for each term. An alternative approach is to approximate the multivariate Poisson or negative binomial processes with the multivariate normal distribution. This allows term frequency dependence to be formally modeled and provides a tractable approximation.

7.9 TERM SEQUENCES AND NON-STATIONARY PROCESSES

Parameters for processes are often treated as though they are *stationary*, that is, the parameters remain constant during the time period under consideration. In some cases, however, a process is non-stationary and one or more parameters change in value over time. In natural language production, for example, as one moves smoothly from one topic to another, one might expect the average frequency of a subject related term to vary smoothly over time. While the terms themselves might be Poisson distributed at one point in time, given a particular parameter, an adequate model of natural language production in such a situation needs to consider the changing parameter values.

One way of addressing non-stationary parameters is to describe the distribution of values and to incorporate this "prior" information about the parameter using Bayesian methods. This method may be useful, albeit flawed, if over the time period of concern the distribution comes close to describing the non-stationary process.

Sometimes parameter values change in a seemingly random manner, yet each value is a function of the previous value or values. This type of pseudo-random function finds frequent use in cryptography, where functions producing numbers that are seemingly random yet can be easily generated are needed for permuting and transposing message characteristics. In other cases, the value of a parameter at a particular time is correlated with (but not the same as) the parameter for previous time periods. For example, there are strong dependencies between data values describing video images, with one image usually being related to the preceding image (Pancha & El Zarki, 1994). A set of images from a night-time scene will contain dark and light images, but will have, on the average, a lower light level than would a comparable scene in daylight. Capturing the nature of these dependencies between parameter values at different periods of time will result in more accurate estimation of probabilities.

In some instances these parameter changes exhibit patterns of change such that the patterns appear the same no matter what the scale of examination. *Self-similarity* is a scale-independent chaotic phenomenon and its implications are only beginning to be appreciated. If a parameter's values are self-similar, capturing this phenomenon is essential, both to accurately estimate probabilities and to explain the results.

Documents or streams of informational text may have relatively constant parameter values over given stretches of the text, with sudden transitions from one state to another. O Ruanaidh and Fitzgerald (1996) suggest methods for determining where these transitions occur, as do Lam et. al. (1996). By detecting *change points*, it is possible to estimate using traditional methods the parameter values over the stretch of a document bracketed by change points. Being able to detect change points is valuable in isolating sections of documents for passage retrieval or portions of network traffic that deal with a certain topic to a certain degree, as well as to estimate the performance associated with retrieving that particular segment of the document.

Markov Models

The relationships between terms, how the occurrence of one term might depend on previous term occurrences, is often described with Hidden Markov Models (HMM). A "regular" Markov Model predicts the probability that one term will be followed by another term (Mandelbrot, 1961). This is represented by a transition matrix

$$\begin{bmatrix} p_{1,1} & p_{1,2} & \cdots & p_{1,c} \\ p_{2,1} & p_{2,2} & \cdots & p_{2,c} \\ & & \cdots & \\ p_{r,1} & p_{r,2} & \cdots & p_{r,c} \end{bmatrix}$$

which has for a given row x and column y the probability $p_{x,y}$ that term x will be followed immediately by term y. These probabilities may be estimated using the general set of *Metropolis-Hastings* estimators (Gelman et al., 1995) which include the Metropolis algorithm (Metropolis & Ulam, 1949) and the Gibbs sampler (Geman & Geman, 1984).

The language generating process may be modeled as a Markov process, in which each term is probabilistically related to the next term. The Markov matrix represents the probability that a given term will be followed by a particular other term. Thus, in a sentence $a\ b\ c$, the occurrence of a partially determines what will occur next (from the domain of possible terms). If our vocabulary is limited to three terms, a, b, and c, the matrix might be something like

$$\begin{bmatrix} .1 & .7 & .2 \\ .5 & .05, & .45 \\ .4 & .5 & .1 \end{bmatrix}.$$

This implies that the probability that term a (first row) will be immediately followed by term c (third column) is .2.

Estimating the probabilities in n-grams becomes increasingly difficult as n grows. This is the general problem with estimating probabilities in higher dimensional spaces. HMMs explicitly address this problem by estimating probabilities through techniques consistent with the sequential nature of language, with one term being generated after another. One can learn different linguistic styles by accurately learning probabilities associated with term and grammatical sequences. Markov Models may be modified so that the transition from one state to another reflects not only the previous state but other values, which may include the states before that (MacKay, 1996; Saul & Jordan, 1995). This is accomplished by hypothesizing *hidden states* that exist between the states represented by the rows and columns in the matrices above. The hidden nodes may be understood as representing the degree to which the nodes at a distance directly influence the probability of the current node. Using HMMs may best be viewed as estimating a probability based on the previous state and a function of the states before that. A variety of learning algorithms have been proposed, with different strengths and weaknesses for the different procedures (Knill &

Young, 1997). Other techniques, such as the use of Boltzmann chains, can produce the same estimates as HMMs and may be used to provide an alternative explanation of process (MacKay, 1996; Saul & Jordan, 1995).

7.10 EMPIRICAL METHODS

Empirical methods may be used to develop probabilistic distributions that are consistent with the data but are otherwise determined by desirable constraints (Maritz, 1970; Fuhr, 1989). The characteristics of such a distribution, like the mean, the variance, or the cumulative distribution function, may be computed using numerical methods. For example, one may develop a distribution consistent with the observed data but that otherwise maximizes the entropy (randomness) of the distribution (Cooper & Huizinga, 1982; Kantor, 1984; Berger, Della Pietra, & Della Pietra, 1996). For a given mean and variance, for example, the entropy is maximized by the normal distribution (Bishop, 1995). Maximizing a function subject to a constraint is often performed by using the method of Lagrange multipliers, discussed in many intermediate calculus texts.

The estimating of complex multivariate probabilities is often performed in Bayesian data analysis by using Monte Carlo simulations of Markov Chains to achieve reliable results (Gelman et al., 1995). Through the use of Gibbs sampling (Dorfman, 1997; Tanner, 1996) one can produce sets of values that can be used to estimate the characteristics of the true distribution. This allows one to estimate parameters of the desired distribution, for example, by taking the appropriate parameters from the sets of values produced during the simulation. Simulations continue until a satisfactory level of convergence is obtained, with most techniques addressing the nature of actual convergence and providing rules to determine stopping points. Limited means are available to analytically determine before starting at what point one can reasonably be expected to stop (Polson, 1996). Simulation techniques such as these have successfully produced accurate parameter estimates in the context of very complex models, such as are found in natural language documents or media that are increasingly being retrieved.

7.11 REDUCING DIMENSIONALITY

Large amounts of dependent data may take so much time to process that using term dependence models may become prohibitive. One way to avoid this is to reduce the dimensions, possibly including the number of terms or concepts, in the data. The *curse of dimensionality,* as it is often referred to, can be minimized through the use of techniques that effectively discard those dependencies that are less important, leaving the more important factors, or one can use techniques that produce statistically independent factors. Many of these methods are based on techniques related to factor analysis, a statistical technique which can locate independent factors that capture the dependencies down to an arbitrarily chosen cutoff. Dimensionality may also be reduced by selecting only terms important to the retrieval process, removing less important terms such as stop-words, e.g. *the, a, an, of, by, for,* and *with* (Schutze, Hull, & Pedersen, 1995).

The most widely discussed method for doing this at present is based upon singular value decomposition (SVD) that is used for Latent Semantic Indexing (Deerwester, Dumais, Furnas, Landauer, & Harshman, 1990). This technique, informally stated, starts with a term-document matrix, breaks it apart into several matrices, removes the less significant portions of the data, and then recombines the matrices, producing a new term-document matrix, with the lesser dependencies removed. In the new document-term matrix, documents are described as linear combinations of uncorrelated terms.

Consider the following matrix derived from the data in Table 7.1. Each column represents a term, and each row a different document:

$$\begin{bmatrix} 0 & 0 & 0 \\ 0 & 0 & 1 \\ 0 & 1 & 1 \\ 1 & 0 & 0 \\ 1 & 0 & 0 \\ 1 & 0 & 1 \\ 1 & 0 & 1 \\ 1 & 1 & 1 \end{bmatrix}.$$

When one applies the SVD procedure, the result is 3 matrices:

$$u = \begin{bmatrix} 0 & -0.236 & -0.339 & -0.220 & -0.220 & -0.456 & -0.456 & -0.560 \\ 0 & -0.301 & -0.619 & 0.472 & 0.472 & 0.170 & 0.170 & -0.147 \\ 0 & 0.571 & -0.282 & -0.210 & -0.210 & 0.361 & 0.361 & -0.493 \end{bmatrix},$$

a diagonal matrix with values

$$m = \begin{bmatrix} 2.946 & & \\ & 1.552 & \\ & & 0.952 \end{bmatrix},$$

and

$$v = \begin{bmatrix} -0.650 & -0.305 & -0.695 \\ 0.732 & -0.493 & -0.467 \\ -0.200 & -0.814 & 0.544 \end{bmatrix}.$$

When we multiply

$$u^T.m.v$$

we arrive at the original document-term matrix

$$\begin{bmatrix} 0 & 0 & 0 \\ 0 & 0 & 1 \\ 0 & 1 & 1 \\ 1 & 0 & 0 \\ 1 & 0 & 0 \\ 1 & 0 & 1 \\ 1 & 0 & 1 \\ 1 & 1 & 1 \end{bmatrix}.$$

If the lowest value in matrix m is replaced with a 0, we arrive at the document term matrix

$$\begin{bmatrix} 0 & 0 & 0 \\ 0.109 & 0.443 & 0.703 \\ -0.0539 & 0.780 & 1.146 \\ 0.959 & -0.163 & 0.109 \\ 0.959 & -0.163 & 0.109 \\ 1.069 & 0.280 & 0.812 \\ 1.069 & 0.280 & 0.812 \\ 0.905 & 0.617 & 1.255 \end{bmatrix}.$$

Here the term values for each document are different than in the original document term matrix, but are not far away from the original values. These values may, in fact, be better representations for retrieval purposes, having discarded possibly extraneous dependencies.

If the two lowest values in matrix m are replaced with 0, then the document term matrix becomes

$$\begin{bmatrix} 0 & 0 & 0 \\ 0.452 & 0.212 & 0.484 \\ 0.650 & 0.305 & 0.696 \\ 0.422 & 0.198 & 0.452 \\ 0.422 & 0.198 & 0.452 \\ 0.874 & 0.411 & 0.936 \\ 0.874 & 0.411 & 0.936 \\ 1.073 & 0.504 & 1.148 \end{bmatrix}.$$

These values are further from the original values, but are still simpler than in the preceding example. While using LSI may result in superior retrieval, the amount of storage needed may be much larger than that used for storing integer frequencies, and inverted files can no longer be used, increasing processing time (Hull, 1994). However, there may be significant functional advantages to using the LSI model. One set of experiments showed that retrieval performance peaks for English language text when the number of independent factors is in the hundreds (Deerwester et al., 1990). Cross language retrieval, where a query in one language can be used to retrieve documents in another language, can be obtained by determining the underlying independent factors in documents written in different languages on the same topic (Littman & Jiang, 1998).

7.12 DISCUSSION AND SUMMARY

Because so many retrieval and filtering systems use multiple features in their queries and virtually all documents have multiple features, the incorporation of multivariate probabilities into the non-experimental determination of retrieval performance is essential. As desirable as this goal is, there is a lack of conjugate priors for many of the most interesting distributions, resulting in the analytic inelegance of multivariate models with many term distributions.

Two simple models of multivariate probabilities may be relatively simply incorporated into non-experimental models. The first, based on the multivariate binary distribution, allows us to consider simple term presence and absence and the relationships between these across a set of terms. The multivariate normal distribution has a relatively easily understood survival function whose use can be understood in a variety of domains, as does the logistic distribution.

In succeeding chapters, these term probabilities and inter-term relationships will be used to model and assess systems addressing other phenomena, such as Boolean logic or natural language processing. The survival based model can also be understood quite nicely if features are understood to have values that are normally distributed. While normally distributed features occur rarely, at least in most existing systems, use of the normal distribution greatly simplifies our survival model and can lead to an increased understanding of the process.

8 PERFORMANCE WITH MULTIPLE TERMS

All hope abandon, ye who enter here.

—Dante's *Inferno*, iii. 9

8.1 INTRODUCTION

A comparison of Figure 8.1 and the smoother and simpler performance graphs found in previous chapters graphically illustrates some of the problems that arise when studying retrieval and filtering performance where more than one term comes into play. Retrieval performance for multivariate techniques that address statistical dependence between document features is poorly understood. Figure 8.1 illustrates the relationship between just a few parameters; the operation of a full system is far more complex. Figure 8.1 shows the performance with two terms and with one thousand documents, with $t = .01$ for both terms, $p = .3$ for the least discriminating term, with p varying for the other term. The correlation between the term frequencies is allowed to vary in relevant documents as shown but is treated as very close to 0 for the set of all documents. It appears as though there are no simple means of exploring this problem, motivating the placement of the quote from Dante at the head of this chapter. The capability to model the performance of these multiple term systems is needed; we develop multi-term models below, including models that explicitly assume that terms are statistically dependent, as well as models as-

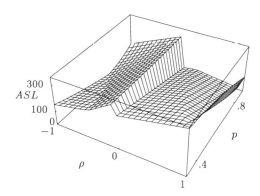

Figure 8.1.   While the first p varies, the second p remains fixed at .3. Also, $t = .01$ for both terms and $N = 1000$. Shown is the correlation between the two features, which is kept the same for both relevant and all documents.

suming multi-term statistical independence. Using an analytic model showing the exact expected performance, we can see the complexity inherent in multi-variate system performance and appreciate why it has been difficult to reach any conclusive experimental results.

One of the most interesting multivariate phenomenon found in filtering is the relationship that exists between natural language terms. The reader is presently making use of a complex system of term relationships, those provided by the grammar of a natural language (Chomsky, 1965; Groenink, 1997). Grammatical rules describe relationships between individual terms, as well as between larger constructs, such as the subject of a sentence and the object of a sentence, where both components may consist of several terms. The ability of a "good" natural language processing system to significantly improve filtering performance has always been assumed; this Holy Grail is a major part of what separates "dumb" existing filtering systems from "smart" humans, who are capable of filtering much better than the "dumb" machine.

In the univariate case, performance is determined by the ranking model, the number of documents in the database, and the probabilistic parameters describing the distribution of all documents in the database, as well as the parameters describing the subset of relevant documents. In the multivariate case, additional factors come into play. Performance with a single term illustrates the tradeoff between two parameters, p and t. When more than one term is used, varying probabilities can result in changes in the relative discrimination values of different terms and rates of change in performance, producing marked increases and decreases in performance when parameters vary, such as the performance found in Figure 8.1. In addition to the effects produced by having additional terms, the statistical interrelationships between terms have an impact on performance.

We use the notation $\mathcal{Q}$ and $\mathcal{A}$ from earlier chapters here when referring to concepts similar to those used with single terms. The variable $\mathcal{Q}$ is used to represent the probability of a ranking in the multi-term case, while in the single term case it was the probability that the ranking was optimal. Similarly, $\mathcal{A}$ is used in the multi-term case to represent the proportion of documents ordered before the expected position of a relevant document for a given ranking, whereas with the single term case it referred to the expected proportion found for optimal ranking.

The models developed in Chapter 6 are analytic in nature, providing symbolic formulae describing system performance. While symbolic solutions are provided here for many multivariate problems, they may be difficult to use or interpret. When studying the effects of larger phrases of 4 or 5 terms, for example, it becomes difficult to make any easily understood general statements about performance. In these cases, we will apply quantitative methods to study system performance.

8.2   ASL

Definition 8.1 (ASL for multiple terms) *The average search length (ASL) or the expected position of a relevant document for multiple terms is*

$$ASL = N \sum_r \Pr(\mathcal{R}_r)\mathcal{A}_r + \frac{1}{2}, \qquad (8.1)$$

where r is taken over all possible rankings, $\mathcal{R}_r$ is ranking r, and $\mathcal{A}_r$ is the $\mathcal{A}$ associated with document ordering r.

An average such as this is minimized when, for the smallest $\mathcal{A}$ value, $\Pr(\mathcal{R}_r)$ approaches 1.

Theorem 8.1 (Minimizing ASL for sub-optimal ranking) *For a set of $\mathcal{A}$ values and a set of probabilities for rankings, the minimum ASL associated with the set of probabilities assigned to the $\mathcal{A}$ values in Equation 8.1 is obtained when the $\mathcal{A}$ values are weakly ordered by increasing value, and the probabilities are weakly ordered by decreasing value.*

The exact system performance may be computed when we have knowledge of the $\mathcal{Q}_i$ ($\mathcal{R}_i$) and $\mathcal{A}_i$ values for all i, where i represents a possible document profile. The average $\mathcal{A}$ is computed over all the possible profiles to produce the ASL as in Equation 8.1.

The *portion* of the $\mathcal{A}$ value contributed by a particular profile d_{i*}, given ranking $\mathcal{R}_r$, is denoted as

$$\mathcal{A}_{ri} = \Pr(d_{i*}|rel) \left(\mathcal{C}_i(D) - \frac{\Pr(d_{i*})}{2} \right).$$

We may compute $\mathcal{A}$ for ranking r as

$$\mathcal{A}_r = \sum_i \Pr(d_i|rel) \left(C_i(D) - \frac{\Pr(d_i)}{2} \right), \qquad (8.2)$$

where the document profiles are weakly ordered by increasing discrimination value. When $\mathcal{A}$ is unsubscripted below, optimal ranking is implied.

Theorem 8.2 (Approximate best-case for multi-term ASL) *When, for some r, the probability of ranking r is 1 and $\mathcal{A}_r$ is 0, the approximate best-case ASL of $1/2$ is obtained.*

Theorem 8.3 (Exact best-case for multi-term ASL) *When, for some r, the probability of ranking r is 1, the lowest value for $\mathcal{A}_r = g/2$, where g is the generality, then the best-case $ASL = N(g/2) + 1/2$.*

Corollary 8.1 (Approximate worst-case for multi-term ASL) *When, for some r, the probability of ranking r is 1 and $\mathcal{A}_r$ is 1, $ASL = N + 1/2$.*

Corollary 8.2 (Exact worst-case for multi-term ASL) *When, for some r, the probability of ranking r is 1, the lowest value for $\mathcal{A}_r = 1 - g/2$, then the exact worst-case $ASL = N(1 - g/2) + 1/2$.*

Corollary 8.3 (Bounds independent of number of terms) *Given the upper and lower bounds computed above for multiple terms and in Chapter 6 for single term queries, the upper and lower case bounds are the same for both single and multi-term cases and the bounds are independent of the number of terms used in discriminating (usually the number of query terms.)*

Example. The ASL may be easily computed for the case in which only one ranking is assumed. Consider the following four document profiles with their associated relevance values:

Profile	Relevance
0 1	relevant
0 1	relevant
1 0	relevant
1 1	non-relevant.

The documents are arranged here in the order in which they will be retrieved. Assume that ranking is optimal, that is, $\Pr(\mathcal{R}_o) = 1$. Using Equation 8.2, we compute

$$ASL = 4 \left(\frac{2}{3}\frac{2}{8} + \frac{1}{3}\frac{5}{8} + 0\frac{7}{8} \right) + \frac{1}{2} = 4\frac{9}{24} + \frac{1}{2} = 2.$$

In this dataset, the $\Pr(d|rel)$ probabilities are $(2/3, 1/3, \text{ and } 0)$, while the cumulative value on the right hand side of Equation 8.2, representing average positions for the document profiles, are $(1/4, 5/8, \text{ and } 7/8)$.

EXERCISE 8.1. [*Easy*] Assume optimal ranking and 4 documents (with binary profiles) and associated relevance values: 11 (relevant), 10 (relevant), 10 (non-relevant), 00 (non-relevant). Compute the ASL using Equation 8.1.

8.3   DIFFERING $\mathcal{A}$ VALUES

It was earlier shown how an $\mathcal{A}$ value may be computed using Equation 8.2. For the case of two terms, one may compute $\mathcal{A}$ based on knowledge about the parameters p_1, p_2, t_1,t_2, the covariances between the term frequencies in relevant documents σ_R, and the covariances between the term frequencies in all documents σ_A.

Covariances

We can make a general argument that when holding other parameters constant, increasing the covariance between the terms will result in an increased probability of the joint distribution. The covariance represents how two variables vary with each other and may be computed as the correlation between the two variables multipled by the standard deviations of the two variables. When this occurs for distributions for relevant and for all documents, and the same covariance is used for both, the discrimination power will increase and $\mathcal{A}$ will decrease.

The relationship between covariances and probabilities may be studied for a simple two-term case as follows. We denote the probability of a particular term pair $p = \{p_1, p_2\}$, $p_1 = p_2$, and the covariance between both variables 1 and 2 is denoted as σ_{12}. We find that

$$\Pr(\boldsymbol{p}, \sigma_{12}) = p_1^{d_1}(1 - p_1)^{1-d_1}p_2^{d_2}(1 - p_2)^{1-d_2}$$
$$\times \left[1 + \frac{\sigma_{12}(d_1 - p_1)(d_2 - p_2)}{p_1(1 - p_1)p_2(1 - p_2)}\right], \quad (8.3)$$

which is derived from the Bahadur Lazarsfeld Expansion, Equation 7.4 (Bahadur, 1961).

For term pairs and two different sets of probabilities and covariances, one primed and one unprimed, the relationship between the two term pairs (with the unprimed parameters being for the performatively-superior $\mathcal{A}$ value) may be characterized as

$$\mathcal{A}' \geq \mathcal{A} \text{ if and only if } \frac{\Pr(\boldsymbol{p}, \sigma)}{\Pr(\boldsymbol{t}, \sigma)} \geq \frac{\Pr(\boldsymbol{p}', \sigma')}{\Pr(\boldsymbol{t}', \sigma')}. \quad (8.4)$$

Equal Probability Values

When the probability vectors on both sides of the inequality are equal for each class of documents, that is, $\boldsymbol{p} = \boldsymbol{p}'$ and $\boldsymbol{t} = \boldsymbol{t}'$, we find that the inequality in Equation 8.4 can be used to study the effects of different covariances.

Theorem 8.4 (Term covariances in relevant documents) *If $\sigma_A = \sigma'_A$, $p = p'$, and $t = t'$, $A \leq A'$ only when $\sigma_R \geq \sigma'_R$.*

Theorem 8.5 (Term covariances in all documents) *If $\sigma_R = \sigma'_R$, $p = p'$, and $t = t'$, $A < A'$ only when $\sigma_A < \sigma'_A$.*

Corollary 8.4 (Ratios of term covariances) *If $p = p'$, and $t = t'$,*

$$A < A' \text{ only when } \frac{\sigma_R}{\sigma_A} \geq \frac{\sigma'_R}{\sigma'_A}.$$

Theorem 8.6 (Dependence vs. independence in relevant documents) *When $\sigma_A = \sigma'_A$, $p = p'$, and $t = t'$, A with a positive covariance between terms in relevant documents is superior to A with term independence ($\sigma_R = 0$.)*

Note that this does not say that retrieval performance with (or assuming) term dependence is always superior to performance with (or assuming) term independence. The A component is the performance when ranking is optimal, and this theorem addresses optimal ranking. This does not address the performance that occurs when there is term dependence and our knowledge about the terms is that they either are or aren't dependent and we don't know the correct parameter values, which are the problems one needs to ask in real world situations where term dependence is a fact of life.

There aren't comparably simple solutions for the situation where the covariances are the same for both A and A' and the probabilities vary.

EXERCISE 8.2. [*Moderate*] Given the discussion in this section, under what conditions would two-term phrases be good discriminators? How would one identify these phrases in practice?

EXERCISE 8.3. [*Research Problem*] Generalize Theorems 8.4 to 8.6 so that they apply to groups of n terms (when $n = 2$, we have term pairs).

1. Try this for $n = 3$ (term triples) and other small n,
2. Try this for $n \to \infty$.

8.4 COMPUTING $\mathcal{Q}$

When one evaluates different ranking procedures, or ranking under different assumptions, it is often required that one compute the probability of a ranking, $\mathcal{Q}_i = \Pr(\mathcal{R}_i)$. This is comparable to computing $\mathcal{Q}$ for the univariate case. Because of the complexity of multivariate models, these may often best be studied numerically rather than symbolically. For example, studying the assumptions made by the Bahadur Lazarsfeld expansion, dependence using a maximum spanning tree, or the generalized term dependence model may be studied for a given degree or type of dependence. Similarly, the effect of using various kinds of phrases and their associated dependence may be studied by noting the difference in $\mathcal{Q}$ for the different phrasings and assumptions.

For multivariate methods that discriminate based on the mean for relevant and all documents, the equations in Chapter 5 such as Equation 5.17 may be used in the multivariate case, with mean vectors replacing the means. One may compute $\Pr(\mathcal{R}_r)$ in Equation 8.1 analytically; however, for several terms, the expression becomes very complex.

Definition 8.2 (Probability of an ordering) *The probability of obtaining ordering $\mathcal{R}_x$, where $x = x_1, x_2, \ldots, x_n$ is a vector of n document profiles ordered as in x,*

$$\Pr(\mathcal{R}_x) = \Pr(w(x_1) \geq w(x_2) \geq \cdots \geq w(x_{n-1}) \geq w(x_n)), \qquad (8.5)$$

where $w(i)$ is the weight for profile i.

Consider the case where there are two document profiles, x_1 and x_2. If the probability distributions describing each profile are peaked so that their variance approaches 0, their weights may be treated as point values. The probability that one profile is greater than the other will approach 1 for one ordering of the profiles. When this peaking occurs with the ordering of profiles associated with the optimal raking, then the best-case $\mathcal{A}$ will be given the highest weight when computing the ASL and the lower quality rankings will be given little weight.

In the case where the variances for the two distributions are high, the probability that one distribution is larger than another will be below 1. When the variances are much higher, and the profile distributions approach the uniform distributions, each ranking may be given similar probabilities, and the $\mathcal{A}$ and $\mathcal{A}$ values will be given nearly the same weight, moving the ASL toward the value obtained with random ranking.

Equation 8.5 may be computed by consider the probability that the first element is greater than or equal to the second, the second greater than or equal to the third, and so forth, combining each of these probabilities. This may be computed as

$$\Pr(\mathcal{R}_x) = \prod_{i=1}^{n-1} \Pr\left(w(x_i) \geq w(x_{i+1})\right). \qquad (8.6)$$

Simple parametric distributions used in an analytic multivariate model may be less likely to capture the essence of the data than is the case with the single term models. The accuracy of parameter estimates drops as the number of terms being considered increases and the number of possible profiles increases. Estimates of the frequencies of occurrence of particular document profiles are based on much smaller training sets, since far smaller sets of documents will have a particular set of variable values. In these cases, the values supplied by parametric estimates are largely determined by the distribution assumed, and this can lead to additional inaccuracies since prior probabilistic distributions are often only approximations of the true random phenomenon being modeled.

Analytic Model

In Chapter 5, Q was derived for a number of univariate cases. These may be generalized by using vectors of parameters rather than individual parameters. For example, Equation 5.17 may be transformed to

$$Q = min\left(S_x\left(h\left(z\right)\right), S_\mu\left(g\left(z\right)\right)\right) + min\left(C_y(h\left(z\right)), C_\mu(g\left(z\right))\right)$$

(8.7)

where the survival and cumulative functions have the z values taken over the given range. The functions $g(z)$ and $h(z)$ in Equation 8.7 represent the actual or true distribution function and the believed distribution function. The parameters x and y represent ranking method specific parameters. For example, when using the decision theoretic model for ranking, $x = \mu_n$ and $y = \mu_n$.

It may prove helpful to note that one can construct a ranking procedure from a formula defining Q like the one above using the same techniques as were used for the single term retrieval model.

One may wish to treat the prior distribution of the mean for a data set as distributed consistent with the multivariate normal distribution. Since the normal distribution itself is the conjugate distribution for the normally distributed population's mean and the inverted Wishart distribution is the conjugate prior for the distribution's covariance matrix, the joint normal-inverted-Wishart distribution is the conjugate prior distribution for the multivariate normal distribution. Using Equation 8.7 and these conjugate distributions, one can study analytic indicators of the effects of various phenomena. Many of the effects described in Chapter 8 can be studied using these distributions.

Historical Method

The probability of a particular ranking may be computed based on historical data. Thus, $\Pr(\mathcal{R}_r)$ is taken from the set of orderings over the set of document profiles. A given ranking occurs, for example, when the first term is ranked above or at the same level as the second, the second at the same level or above the third, and so forth. The ranking $\{11, 10, 01, 00\}$ occurs when

$$\frac{p_{11}}{t_{11}} \geq \frac{p_{10}}{t_{10}} \geq \frac{p_{01}}{t_{01}} \geq \frac{p_{00}}{t_{00}}.$$

While it may be very probable that the first document has a ratio exceeding the second, and a high probability that the second ratio is greater than or equal to the third, and so forth, the probability that all these will occur together is substantially lower than the probability that one profile alone is greater than or equal to another profile. The probabilities may be tabulated when evaluating each of the profile pairings that may be considered when ranking.

Best-worst Model

Estimating $\mathcal{A}$ or $\mathcal{Q}$ values may be difficult in the multivariate case because of the large number of cases to be considered. It would be useful to estimate the ASL from the $\mathcal{Q}$ weighted average of $\mathcal{A}$ by considering just the best and worst case $\mathcal{A}$ values, rather than the full range of $\mathcal{A}$ values for each possible ordering. One might wish to accept

Conjecture 8.1 (Best-worst case estimation) *The ASL as estimated in Equation 8.1 may be estimated satisfactorily by weighting the $\mathcal{A}$ values for the best (B) and worst (W) cases only with $\mathcal{Q}_B$ and $\mathcal{Q}_W$. Thus,*

$$ASL \approx N \left(\frac{\mathcal{A}_B \Pr(\mathcal{R}_B) + \mathcal{A}_W \Pr(\mathcal{R}_W)}{\Pr(\mathcal{R}_B) + \Pr(\mathcal{R}_W)} \right). \tag{8.8}$$

This is a good approximation when the distributions for $\mathcal{A}$ and $\mathcal{Q}$ are close to symmetrical This conjecture provides a useful tool for comparing different methods, with many problems with the conjecture not affecting the value of a comparison.

Random Selection of Profiles

One may also estimate the *ASL* by randomly sampling from the set of profiles and then estimating the $\mathcal{Q}$ from the sample pairs that are drawn. The probability of a ranking being optimal is estimated by taking pairs of different profiles and then determining whether the rankings $\mathcal{R}_x$ and $\mathcal{R}_o$ are the same for these pairs. By doing this repeatedly, one can estimate how often the rankings are the same. In some circumstances, one needs the probability of a ranking rather than the probability of optimality, that is, the probability of the ranking that is the same as the optimal ranking. The probability of a given ranking being obtained, as in Equation 8.1, requires a large sample from which to estimate the probability with any accuracy. These random estimators of $\mathcal{Q}$ may use any of a number of techniques, some of which are best for larger samples, and some of which may be used for very small samples such as those found in relevance feedback systems (Gelman et al., 1995).

8.5 UNDERSTANDING RANKING WITH TERM DEPENDENCIES

Since the performance of filtering systems depends on the ordering of documents with different profiles, a closer examination of the probabilities of the possible document profiles and the resulting document rankings is in order. Consider that for two binary terms, there are four possible profiles:

$$
\begin{aligned}
&11\\
&10\\
&01\\
&00.
\end{aligned}
$$

They are ordered this way if the following constraints are used:

1. Binary term frequencies are re-parameterized, if necessary, so that both terms are positive discriminators. Thus, we will be able to assume that $p/t \geq (1-p)/(1-t)$ for each term.

2. The leftmost term has a greater discrimination power than does the term to its right. Thus, using subscripts l and r to denote left and right, $p_l/t_l \geq p_r/t_r$.

These two constraints are notational, and any pair of terms can meet them if the terms are re-parameterized and reordered, if necessary.

Remembering that the probabilities for term pairs in documents are computed as follows:

$$
\begin{aligned}
p_{11} &= p_1 p_2 + \sigma_{12} \\
p_{10} &= (1-p_1)p_2 - \sigma_{12} \\
p_{01} &= p_1(1-p_2) - \sigma_{12} \\
p_{00} &= (1-p_1)(1-p_2) + \sigma_{12}
\end{aligned}
$$

when p_i is the probability for term i and σ_{ij} represents the covariance between term i and j. When the correlation coefficient is 0 (and thus the covariance σ_{12} is also 0), the ordering is as with the two constraints above, i.e., $\{11, 10, 01, 00\}$.

Increasing Positive Association

Consider what happens with a slight increase (starting from 0) in the covariance for relevant documents. At the top of the set of ranked document profiles, the probability for 11 increases while the probability for the next profile, 10, decreases, with the increasing covariance added to 11 and subtracted from 10. On the bottom end of the profiles, the probability of 01 decreases and the probability of profile 00 increases, with the two probabilities passing each other at the break-even point where the correlation is:

$$
p_{01} = p_{00} \equiv \rho_{1,2} = \frac{(p_2 - 1)(1 - 2p_1)}{2\sqrt{(p_1 - 1)p_1(p_2 - 1)p_2}}. \tag{8.9}
$$

As the correlation and covariance further increase, the probability of document profile 00 may increase to eventually become larger than the probability for profile 10. The probabilities p_{00} and p_{10} are the same under perfect positive or negative correlations when $p_1 = (1 - \sqrt{p_2})/2$.

Decreasing (Negative) Association

Conversely, as the covariance decreases and the correlation moves downward from 0, the ordering of profiles will change as the probability of 11 drops (at the negative of the values computed above), possibly to below 10. The break even point, where documents with profile 11 are ranked the same as documents with profile 10, is at

$$p_{11} = p_{10} \equiv p_{12} = \frac{p_2(1 - 2p_1)}{2\sqrt{(-1 + p_1)p_1(-1 + p_2)p_2}}. \tag{8.10}$$

EXERCISE 8.4. [*Moderate*] Given the break-even $p_{1,2}$ as computed in Equation 8.9, what is $\sigma_{1,2}$? What will be the numeric value for the break-even $p_{1,2}$ if $p_1 = 4/5$ and $p_2 = 3/4$?

$\mathcal{A}$ for Two Term Models

If the ordering constraints from above are accepted so that terms are positive discriminators moving from the conceptual high order position on the left to the low order position on the right, one may compute $\mathcal{A}$ for term pairs as

$$\mathcal{A}_2 = \Big(p_{11}t_{11} + p_{10}(t_{10} + 2t_{11})$$
$$+ p_{01}(t_{01} + 2(t_{10} + t_{11})) + p_{00}(t_{00} + 2(t_{01} + t_{10} + t_{11}))\Big)/2.$$

If we wish to assume that the terms are independent, this becomes

$$\mathcal{A}_{2,ind} = (1 + p_2(t_1 - 1) + t_1 + t_2 - t_1t_2 + p_1(p_2 - 2p_2t_1 - t_2 + 2t_1t_2 - 1))/2,$$

where p_i represents the probability of term i occurring in relevant documents, with corresponding meanings for t_i.

Degree of Dependence and Discrimination

It may be helpful to formalize the relationship between the degree of dependence between n terms and the amount of discrimination provided by incorporating that degree of dependence. We suggest the following:

Conjecture 8.2 (Dependence — discrimination relationship) *The additional contribution to discrimination due to incorporating dependence of degree n (i.e., n-way dependencies) is greater, on the average, than the additional contribution made by incorporating dependence of degree $n + 1$. The expected contribution for all n is positive.*

When one has a given query term, the further one moves in the query away from occurrences of this term, the less likely the terms are to be related to the query term. Part of this is due to the relationship between term distance and dissimilarity. In other circumstances, the added terms are increasingly likely to be from a different sentence or phrase and be on a different topic as the amount of dependence information increases.

There is empirical support from studies of increasing degrees of term dependence (Losee, 1994a) and varying sizes of sets of contiguous terms (windows) (Haas & Losee, 1994; Jacquemin, 1996) that increasing the number of neighboring terms included in document weighting generally increases retrieval performance, although the rate of increase changes more at certain window sizes and the rate decreases after a certain point.

Assuming that distance between two terms may be defined as the number of terms between the terms in question, or the number of grammatical units between the two terms, these phenomena suggest the following:

Conjecture 8.3 (Distance — discrimination relationship) *Given term x, and any term y at a distance δ from x, the average ability of all terms y at a distance δ to discriminate between relevant and non-relevant documents given a query about topic x will decrease as δ increases.*

Conjecture 8.4 (Distance — association relationship) *Given a term x, and all terms y at a distance δ from x, the average correlation between terms x and y at a distance δ will decrease as δ increases.*

EXERCISE 8.5. [*Moderate*] Design experimental studies that would provide support for Conjectures 8.2 to 8.4.

8.6 A CASE STUDY

The usefulness of a term dependence model can be seen by studying its application to a document database. As an example, consider the query *text information retrieval* applied to the set of book titles in Table 8.1. From the 22 titles and using the matrix models of term dependence described in the previous chapter, we can compute the following probabilities for individual terms

Term	p	t
retrieval	3/22	6/22
information	3/22	9/22
text	2/22	7/22

with the probabilities for individual profiles

Relevance	000	001	010	011	100	101	110	111
Rel		1/4					1/2	1/4
All	7/22	4/22	5/22			2/22	3/22	1/22

The term *retrieval* is clearly the best discriminator, and is thus placed on the left in Table 8.1, with *information* being the next best discriminator and *text* being the weakest discriminator and thus being on the right.

We can compute the covariance vector for this data set of relevant documents as

$$\sigma_{rel}^{(3)} = [1, 0, 0, -1/8, 0, -1/8, 3/16, 1/16].$$

Similarly, the covariance vector for all documents is

$$\sigma_{all}^{(3)} = [1, 0, 0, -41/484, 0, 6/121, 17/242, -51/5324].$$

Table 8.1.   Twenty-two sample titles and four arbitrary relevance judgements for the query "Text Information Retrieval," with the terms listed in the table in order of decreasing discrimination value.

Term Presence				
Retrieval	Info.	Text	Rel	Title
1	1	0	R	Information Retrieval
0	0	1	R	Automatic Text Processing
1	1	0	R	Introduction to Modern Information Retrieval
1	1	1	R	Text Information Retrieval Systems
1	1	0	N	Concepts of Information Retrieval
1	0	1	N	Text Based Retrieval
0	0	0	N	The Illusion of Reality
0	0	0	N	Sphere Packings, Lattices, and Groups
0	1	0	N	Entropy, Information, and Evolution
0	1	0	N	Measuring Information
0	0	1	N	Text Algorithms
0	1	0	N	Information Theory and Reliable Communication
0	1	0	N	The Science of Information
0	1	0	N	Key Papers in the Economics of Information
0	0	0	N	Optimal Statistical Decisions
0	0	0	N	Laws of the Game
0	0	0	N	Bayesian Reliability Analysis
0	0	0	N	The Age of Electronic Messages
0	0	0	N	The Cuckoo's Egg
0	0	1	N	Text Databases
0	0	1	N	Text Based Intelligent Systems
1	0	1	N	Text Retrieval
3	3	2		*Number in relevant documents*
6	9	7		*Number in all documents*

As one would expect, there is a positive covariance between all three query terms in relevant documents, while over the database as a whole, there is a negative three-way covariance between all the three terms. For the term pair *information – text* we find a negative covariance for relevant documents, while there is a positive covariance for all documents. Examining the relevant documents suggests why; in most relevant documents, when one of the terms *information* or *text* occurs, the other is absent.

The A value for optimal ordering is .255 and the ASL is 5.625.

EXERCISE 8.6. [*Moderate*] Select 10 of your favorite book titles and make a multi-term query for which at least two of the books are relevant. Compute A and the ASL for this query, assuming optimal ranking.

8.7   PERFORMANCE ASSUMING BINARY INDEPENDENCE

Multivariate retrieval models may be simplified when terms are assumed to be statistically independent. Independent terms provide no information about each other occurring, thus, knowing that the term *filtering* occurs in a document makes it no more and no less likely that the term *text* occurs.

Informally, the method used here to compute performance is to stack a series of $\mathcal{A}$ values, each of which is scaled downwards, so that when all the $\mathcal{A}$ values are stacked together and combined, using the $\mathcal{Q}$ value for the appropriate terms, they can take on the values from 1 to 0. Term independence allows us to compute this in a relatively straightforward manner. The $\mathcal{A}$ values for individual terms are combined into $\mathcal{A}$ values for term pairs, then triples, and so forth, until values are computed for the entire set of possible documents.

The Recursive Computation of ASL

One method of computing the ASL assuming term independence is recursive in nature. Consider Equation 8.1 and how it might be applied in a manner that is consistent with the assumption of statistical independence.

A valuable tool in studying performance under independence is the reindexing of documents. The document profiles can be modified so that when the features are ordered from the best discriminating term to the worst, each discriminator is greater than the sum of all the lesser discriminators. The value for the document weight, the value for the overall number, remains the same; the way of representing it has changed. We can say that each term *dominates* terms with lower discrimination values (Losee, 1991).

When each term dominates the terms with lower discrimination values, one can move through the set of different document profiles by *counting*, with the greatest discriminator being in the most significant digit position and the lowest discriminator being the least significant digit. Counting here is understood to be the process whereby each position in the profile represents a value in a number system, and one can count thorough a number in the number system by moving through all the possible values in the rightmost (least significant) digit, then moving to the next digit to the left, moving through all the values until moving to the next digit position to the left, and so forth.

Term dominance may be obtained through the process of *reindexing*, where a set of index terms are produced that will meet the requirements of term dominance, and then the documents are reindexing using the new terms.

A Recursive Measure

Using the assumptions of binary term distributions and term independence, we may compute the document filtering performance, as measured by the ASL. This technique places increasing limits on the breadth of the segment examined, with the segments decreasing in size as the depth of the recursion increases. The segments in document vector d are delimited by the lower limit ℓ and upper limit h:

$$f(\boldsymbol{d}, i, \ell, h) = \Big(\mathcal{Q}_i [f(\boldsymbol{d}, i+1, \ell, Pr(d_i)(h-\ell)+\ell)$$
$$+ f(\underset{d_i \to \overline{d}_i}{\boldsymbol{d}}, i+1, \Pr(\overline{d}_i)(h-\ell)+\ell, h)]$$
$$+ (1-\mathcal{Q}_i)[f(\underset{d_i \to \overline{d}_i}{\boldsymbol{d}}, i+1, \ell, \Pr(\overline{d}_i)(h-\ell)+\ell)$$
$$+ f(\boldsymbol{d}, i+1, \Pr(d_i)(h-\ell)+\ell, h))] \Big)$$

$$\text{when } i \leq n \tag{8.11}$$

$$f(\boldsymbol{d}, i, \ell, h) = \Pr(\boldsymbol{d}|rel)((h-\ell)/2+\ell) \quad \text{when } i > n. \tag{8.12}$$

This algorithm takes the mean position of a relevant document for each term, and then combines it with the information for the term to its left, in the next most significant position. The $\mathcal{A}$ value is first found for each pair of profiles that differ only by the least significant digit (LSD). The positions are combined with $\mathcal{Q}$ values for the LSD. The processing of each pair is performed by one application of Equation 8.11, which will require multiple executions of Equation 8.12. The two $f()$ functions being added together can be understood as computing the portion of Figure 6.1 for both $d = 1$ and for $d = 0$. Each $\mathcal{Q}$ component, $\mathcal{Q}$ and $\overline{\mathcal{Q}}$ for term i, is used to call the two different $\mathcal{A}$ orderings, the forward and the backward ordering.

8.8   DIFFERENT DEPENDENCE ASSUMPTIONS

The $\mathcal{A}$ value is computed based upon the optimal ranking. Even under the case where the designer, system, or user makes bad assumptions or operates in a state of full or partial ignorance, $\mathcal{A}$ has been assumed to be based on the optimal ranking. The $\mathcal{Q}$ values, on the other hand, explicitly address both the optimal ranking and the (usually sub-optimal) ranking that is used by the retrieval system, based on whatever knowledge is available to the ranking procedure.

It is often helpful to denote the performance obtained with various levels of dependence being significant or being believed to be significant. The dependence that *exists* is represented by the superscript to the left of $\mathcal{P}$ and the degree of dependence *believed* to be significant being represented by the subscript to the left. Thus, $_1^3\mathcal{P}(x)$ represents the performance when three way dependencies are significant when the user believes terms to be independent, that is, dependent to degree 1. By belief, we mean assumed knowledge with which the system is made consistent. In those cases where we wish to represent different degrees of dependence existing or believed to exist in relevant or in all documents, this is represented as a pair of values in the order {relevant, all}. Thus, if we wished to represent actual dependence to degrees 3 and 3 for relevant and all documents and the user believes there is independence and pairwise dependence for relevant and for all documents, respectively, this is denoted as $_{\{1,2\}}^{\{3,3\}}\mathcal{P}(x)$.

Separating the relevant and all sets of documents when examining dependence can prove useful since we can often make claims about all documents because anyone can easily compute these probabilities before searching begins. Parameters of relevant documents can not be precomputed because we do not know the exact nature of the set of relevant documents until after the relevant documents have all been retrieved. Thus, a system designer is likely to assume perfect knowledge about all documents while assuming little knowledge about the set of relevant documents, and our models need to be able to reflect this. Using notation such as this, we can describe the performance obtained with different models. For example, we might use the Bahadur-Lazarsfeld Expansion to degree 3 to estimate the parameters describing all the documents, while we might assume the binary independence model when estimating the characteristics of relevant documents. In addition, fractional degrees of dependence are used in some models, where, for example, the most important pairwise relationships are included but not all pairwise relationships.

EXERCISE 8.7. [*Research Problem*] Consider a two term query with documents ranked by decision-theoretic weighting. Assume that two way dependence holds but the ranking method only uses single terms (term independence) when computing probabilities. Analytically, what is the performance obtained assuming (incorrectly) that only individual term probabilities are needed, as compared to the performance assuming 2 way dependencies are known? More generally, what would be the difference in performance with n-way dependencies but where the system assumes or uses only $n-1$ way dependencies? More generally, assuming that we can pre-compute dependencies for all features easily, what is the performance $_{\{y,n\}}^{\{x,n\}}\mathcal{P}(x)$ where $n \to \infty$?

8.9   QUERY LENGTH AND EXPANSION

Retrieval with multiple terms has obvious advantages over retrieval with only one term, but when should one stop adding terms in an effort to improve performance? The performance with a query of a given length is a function of the probability that a given ranking $\mathcal{R}_x$ occurs, $\Pr(\mathcal{R}_x)$, and the $\mathcal{A}$ value associated with that ordering, $\mathcal{A}_x$. Upper bound retrieval for a given query length occurs when, for some ordering x, $\Pr(\mathcal{R}_x)$ is maximized and $\mathcal{A}_x$ is minimized. Clearly, when retrieval is optimal, $\mathcal{A}_x$ is less than or equal to any other $\mathcal{A}_y$, $x \neq y$, and $\Pr(\mathcal{R}_x) = 1$.

Theorem 8.7 (Adding a term) *Adding a new and positively discriminating term to a query under optimal ranking always results in improved performance.*

If a positively discriminating term is added to an optimal system, $\Pr(\mathcal{R}_x)$ will remain at 1 and $\mathcal{A}_x$ will decrease, because the term is an additional positive discriminator (if it is not the same as a term already in the query).

Adding a term to a query may have a different interpretation in a Bayesian context. Each possible term in the domain of terms may be interpreted as being in the original query, with the vast majority of terms having no discrimination power. What some refer to as query expansion is essentially term reweighting for a few terms. Thus, adding a term may be viewed as re-weighting that term. Relevance feedback similarly reweights query terms.

Theorem 8.8 (Adding a term under sub-optimal ranking) *A positively discriminating term added to a query under sub-optimal ranking will improve performance if and only if the decrease in $\mathcal{A}$ for the better rankings will more than compensate for the decrease in the probabilities for the better rankings.*

The decrease in the probabilities of the rankings is due, in part, to increased ignorance about the characteristics of relevant documents because of the increased number of characteristics. As was seen in Chapter 5, Q usually decreases when there is less knowledge about probabilistic parameters. Note that in many relevance feedback and query expansion situations, one has less knowledge about the "added" terms than is needed to have them produce an improved level of performance.

ASL improvement due to one added term

Consider the ASL computed as with Equations 8.1 and 8.2, where we have the original probabilities and $\mathcal{A}$ values before adding a term, and similar variables associated with adding the new term. The original components have a single subscript which represents the original ranking number. The values associated with the added terms have two subscripts. The first subscript represents an original ranking number and the second subscript the possible values for the term from the m possible values that the query might take on because of the added term, or the size of the domain of values for that term. With the original (unexpanded) query, the portion of the ASL due to a given ranking r is computed using $\Pr(d|rel)$, which we denote as P, and a component representing the proportional cumulative distance to that point in the search (similar to $\mathcal{A}$), which we denote as C. Before adding the term, the portion of the ASL for a given ordering r may be approximated by the product PC, while with an added term with m possible values, each original ranking may be transformed into m new and different ranking. Over the n different original profiles, the difference in performance for a given ranking r is

$$\Delta_r = \sum_{i=1}^{n} \left(\sum_{j=1}^{m} P_{i,j} C_{i,j} \right) - P_i C_i,$$

where all values are those from ranking r. This equation reflects the difference between the average PC value with the added term and the original PC product from the unexpanded query, taken over all n original ranking.

Theorem 8.9 (ASL improvement with an added term) *The differences in performance associated with each original ranking are averaged over the set of all possible rankings to produce*

$$\Delta_{ASL} = \sum_r \Pr(\mathcal{R}_r)\Delta_r. \tag{8.13}$$

A positive amount here represents the amount by which the ASL increases (worsens) when adding a query term, while a negative amount represents the amount by which it decreases (improves.) When a term is added to a query, performance is improved if and only if the decrease in the A for specific rankings outweighs the decreasingly small probabilities of these rankings.

Corollary 8.5 (ASL change under optimal ranking) *Given optimal ranking, when adding a term we may compute*

$$\Delta_{ASL} = \sum_{i=1}^{n} \left(\sum_{j=1}^{m} P_{i,j}C_{i,j} \right) - P_i C_i \tag{8.14}$$

where P and C are for optimal ranking.

Example of Query Expansion

Consider a case under optimal ranking where we have a single term query and then a second term is added to the query. Earlier in this chapter, an example was given with 4 documents that we can use to study query expansion: two relevant documents had profile 01, one relevant document had profile 10 and a non-relevant document had profile 11. Assume that we begin with the original query containing only the first (leftmost) term. The ASL when using only this single term will be 2.167.

When the right term is added, we know from the earlier example that the $ASL = 2$. This may be computed from the single term $ASL = 2.167$ by adding $\Delta_{ASL} = -1/6$ to produce $ASL = 2$. We compute Δ_{ASL} using Equation 8.14, [with profiles provided in square brackets] as

$$\left(\left(0 \ [00] + \frac{2}{3}\frac{3}{2}[01] \right) - \frac{2}{3}\frac{3}{2} \right) + \left(\left(\frac{1}{3} \ 3 \ [10] + 0 \ 4 \ [11] \right) - \frac{1}{3}\frac{7}{2} \right).$$

As can be seen in the second part of this expression, the ASL will drop by the difference between $(1/3)3 = 6/6$ and $(1/3)(7/2) = 7/6$, or by $1/6$.

EXERCISE 8.8. [*Research Problem*] What, analytically, is the optimal number of terms that should be included in a query given optimal ranking? Given sub-optimal ranking?

8.10 BROWSING

When browsing through a set of relevant documents, the number of documents one is expected to pass by is the *expected browsing distance* (EBD). If one begins in the middle of a group of r documents with identical profiles, then the EBD for this group will be $r/2$. For more than one group it is the distance from the expected position of a relevant document on the left to the expected position of a relevant document on the right. More generally, when documents are ordered by the Gray code, counting up $+$ and down $-$ from the starting points,

$$\text{EBD} = N \sum_{ud=+,-} \sum_{i} \Pr(D_i|rel) \left(\frac{h(i) - \ell(i)}{2} + \ell(i) \right) \qquad (8.15)$$

The group of documents with profiles D_i are Gray coded values and are incremented $(+)$ or decremented $(-)$ depending on the ud (up-down) variable. The distance from the starting point, presumed to be the middle of a group of documents, to the start of group i is $\ell(i)$, $\ell(i) = 0$ for the starting group, and the distance from the starting point to the furthest distance of group i is denoted $h(i)$. Both distances have as their maximum the half-way point around the ring of documents.

8.11 SUMMARY

Filtering and retrieval performance for multi-term queries may be modeled with analytic methods. The average search length (ASL), the expected position of a relevant document, may be computed using Equation 8.1. This uses the computation of $\mathcal{A}$, as in Equation 8.2, and $\mathcal{Q}$, as in Equation 8.7.

Performance in the multivariate case is a function of both the individual term parameters and covariance or correlation values describing the statistical relationships between the terms. The effect of these correlations on performance has been examined and, using these methods, we can compute the exact effect of a given correlation. One can compute multivariate binary probabilities using either the matrix based methods in Chapter 7 or using Equations like 8.12.

The relationship between the probability of a ranking and the $\mathcal{A}$ value affects performance directly. In situations such as increasing the length of a query by adding terms, we find that the optimal $\mathcal{A}$ value will improve while the probability of achieving optimal $\mathcal{A}$ will decrease.

9 LOGICS AND RULES

Nothing is as practical as good theory.

—*Harpers*, December 1993

9.1 INTRODUCTION

Logic-based systems have long been used in the retrieval of documents and bibliographic records. The performance of early manual systems, and more recently automated retrieval systems, can be determined using the analytic methods described above. These inherently quantitative methods may be applied, in many cases, to logical queries because of the set-theoretic basis for logic and for probability theory. Since we can determine the size of sets associated with variables referred to in logical statements, we can compute the frequencies and probabilities needed for estimating performance. In some cases logic based retrieval will produce search results for the user inferior to results produced by a system using the probabilistic model that allows for varying probabilities of relevance. However, logical systems are relatively easy to implement and may retrieve documents faster than probabilistic systems, making logic-based systems useful in many situations.

Propositional logic and its variants can serve as the basis for reasoning systems that serve as the basis for many filtering and retrieval systems (Lalmas, 1998; Sebastiani, 1998). These systems are found in search engines that allow for simple **and**ing and **or**ing of query terms, expressing a relationship that needs to exist among terms in the set of retrieved documents. A query such as *scalp and lice,* when used to retrieve documents on head lice and their eradication, is an expression in Boolean logic. Other forms of logic are useful and can introduce more advanced concepts and capabilities into inferential and deductive systems, although logics not based on set theory may not be amenable to analytic modeling.

Logical reasoning is the process by which deductions are made, suggesting one or more *conclusions.* A logical *argument* for a particular conclusion may be a set of statements or propositions, one or more of which are *premises* and one or more of which are conclusions that can be deduced from the premises. Propositions are statements that need not be believed to analyze the validity or effect of an argument. The arguments are normative, that is, they attempt to show how an argument should take place and not how it actually takes place in a natural language. Effective arguments may be made using either deductive reasoning or inferential reasoning.

The statement that a particular argument is valid may be true or false. One may consider an argument "good" when the argument is both valid and the premises are true (Sainsbury, 1991). An argument itself may not be true or false.

Several common logical forms may be used to produce conclusions or statements from other sets of statements. These may be viewed as forms of argumentation by a rhetoritician or as derivational rules by a mathematician. A syllogism is a set of statements containing some statements that are accepted and other statements that follow from these of necessity. (Kneale & Kneale, 1984). This can be viewed as an "if ... then ... " statement, with the premises being conjoined and with a single conclusion. The premises are sometimes referred to as the antecedent and the conclusion or result as the consequent. A sample syllogism is as follows:

> all humans are mortal
> all Greeks are humans
> therefore, all Greeks are mortal.

This form of argument contains set theoretic relationships, that is, Greeks are members of the set humans, which, in turn, have the attribute "mortal."

A non-set theoretic deduction might be as follows:

> if it rains (r) the ground becomes wet (w)
> It is raining
> therefore the ground is wet.

This may be expressed symbolically as $(r \supset w)$, r, therefore w. The operators are those defined earlier on page 57, and the "$\supset$" operator represents "implication." This form of rule is sometimes referred to as *modus ponens.*

Consider another form of argument:

> if it rains (r) the ground becomes wet (w)
> The ground isn't wet
> therefore it isn't raining.

This is an example of another rule, *modus tollens*, which may be symbolically represented as $(r \supset w)$, $\neg w$, therefore $\neg r$.

The relationship between conjunction and disjunction operation is characterized by *DeMorgan's Laws*:

$$\neg(p \wedge q) \equiv (\neg p \vee \neg q)$$
$$\neg(p \vee q) \equiv (\neg p \wedge \neg q).$$

If one wants to prove $p \supset q$, one may use the *reductio ad absurdum* method by assuming both p and $\neg q$ and then looking for a contradiction which would prove that the premises are inconsistent. If $p \supset q$ is false, no contradiction will be obtained.

Predicate Logic

Assume that the predicate Ax represents the fact that document x is about airplanes and Wx represents the fact that document x is about the construction of wings for supersonic aircraft. One might conclude that $Wx \supset Ax$, that is, if a document is about wings for supersonic aircraft then it is about airplanes. *Predicates* such as these are usually written with capital letters and the *arguments* are in lowercase characters to the right of the predicate, e.g., Px. In first-order predicate logic, the arguments are limited to individual items, and not to sets of items or classes of items.

The choice of predicates used in a large deductive system needs to be considered carefully. In the aviation documents example, it might be more efficient to have an "about" predicate, Bxy, where the first lowercase character represents a document identifier and the second lowercase character represents the subject of the document. That document (d) is about "wings" (w) might be represented as Bdw. The earlier expression $Wx \supset Ax$, may be re-expressed as $Bxw \supset Bxa$, if book x is about wings, then book x is about aircraft.

Several predicates are frequently used in reasoning systems (Rowe, 1988). Items may be of a particular type, such as a Boeing 777. They may represent certain properties, such as having gray paint. Predicates may indicate relationships between items, such as the relationship between an airplane and its manufacturer.

Some predicates are quantitative, including numeric functions. Subtraction, for example, may be seen as a predicate. Predicates may also have other predicates as arguments. For example, one predicate might indicate the probability that a second predicate is true or correct.

Quantification

In many instances, one can make additional logical statements if the distinction can be made between predicates that are true for all arguments and those predicates that are true for at least one argument. If Px is true for all xs, this may be formalized as $(\forall x)(Px)$ or, in other common forms, $\forall x Px$ or $(x)Px$. We may read $\forall x$ or (x) preceding an expression as *for all x*. The fact that there is at least one x extant such that Px may be formalized as $(\exists x)Px$. We may read $\exists x$ as *there exists an x such that*. It is the case that

$$(\forall x)Px \quad \supset \quad (\exists x)Px$$
$$(\exists x)\neg Px \quad \supset \quad \neg(\forall x)Px$$
$$\neg(\exists x)Px \quad \supset \quad (\forall x)\neg Px.$$

We can comfortably say, although possibly with some regret, that all humans (H) are mortal (M). Not all humans are female (F), however, some humans are female. This may be formalized as $(\forall x)(Hx \supset Mx)$, that is, for all x's, if x is human then x is mortal. Similarly, $(\exists x)(Hx \supset Fx)$ or $(\exists x)Fx$, that is, there is at least one x such that, if they are human, then they are female.

The earlier expression $Wx \supset Ax$ may now be more precisely expressed as $(\forall x)Bxw \supset Bxa$, that is, for all x's, if a document is about wings than that same document addresses airplanes.

For purposes below, it may prove helpful to express universal and existential quantification in terms of simpler operations. Universal quantification may be understood as an infinite conjunction of a series of predicates for all the universally quantified values. Thus, $\forall x Px$ might be expressed for the n possible values of x as

$$\forall x Px = Px_1 \wedge Px_2 \wedge \cdots \wedge Px_{n-1} \wedge Px_n. \tag{9.1}$$

This says that Px is true for all the values of x, no matter which x is used.

Existential quantification may similarly be understood as the infinite disjunction of the predicate being considered, or

$$\exists x Px = Px_1 \vee Px_2 \vee \cdots \vee Px_{n-1} \vee Px_n. \tag{9.2}$$

Here we claim that the Px_i predicate is true for some i.

The probabilities associated with quantified statements may be computed exactly in some cases, while in other cases the values are limits on the probabilities. Given the statement $\forall x Px$, it is the case that all Px will be true, while the probability that a particular predicate Py is true from among all predicates Pz may be simply computed. An existentially quantified statement such as $\exists x Px$, shows that there exists at least one Px. This can be construed as placing a bottom limit on the probability of Px.

EXERCISE 9.1. [*Easy*] Give some examples of different kinds of natural language statements that can't be represented adequately using one of these logics.

9.2 CONDITIONAL STATEMENTS

Entailment, or logical causation, is one of the cornerstones of reason (Anderson & Belnap, 1975; Diaz, 1981; Sanford, 1989). It implies that a relationship exists, possibly a necessary relationship, between the the two statements or arguments. Given two statements, "Bob is happy today" and "if Bob is happy today then Lee Ann is happy today," one can conclude that "Lee Ann is happy today" is a true statement This form of implication is interpreted as

$$p \supset q \equiv \neg(p \wedge \neg q) \equiv \neg p \vee q.$$

Thus, the implication $p \supset q$ is true except when p is true and q is false. Philo of Megara first defined what has come to be called *material implication*, where a conditional expression is true if the antecedent is false or the consequent is true. G.E. Moore explains entailment as the converse of "is deducible from" (Sanford, 1989). Georg Henrik von Wright (1957) claims that

> p entails q, if and only if, by means of logic, it is possible to come to know the truth of $p \supset q$ without coming to know the falsehood of p or the truth of q.

It is important not to confuse use and mention: Bertrand Russell notes that "propositions are mentioned rather than used" (Sanford, 1989). Truth is assigned to statements, not to the relationship described in the statement. For example, while the statement that "Bob is hungry" may be true or false, the actual state of nature whereby Bob either is or is not hungry does not itself have a truth value. Russell says

> The essential property that we require of implication is this: 'What is implied by a true proposition is true.' It is in virtue of this property that implication yields proofs (Russell, 1906).

Quine similarly stresses that implication is a relation between the names of statements, not between the statements (Sanford, 1989). The statement p implies q provides the names of two statements, the statements identified by p and by q. Also implication is not between objects. Quine claimed (Sanford, 1989) there is not much relationship between $\supset$ and "implies."

Because of the ambiguity associated with implication, it is best understood as a relationship between statements and not as a "connective" like **and** and **or**.

Paradoxes of Material Implication

Material implication $(p \supset q)$ may be represented as "either p is false or q is true." If $p \supset q$, then $\neg p$ can imply that q is true. Thus, given $p \supset q$, any $\neg p$ implies q or $\neg q$. A false proposition in the antecedent, merely because it is false, allows one to imply every possible proposition. Note that to properly understand this, one may need to examine and consider the logical expressions closely, as the common English language interpretations for some of these expressions may be misleading.

Bertrand Russell notes that "of any two propositions there must be one which implies the other." Consider the following progression:

$$(p \supset q) \quad \lor \quad (q \supset p)$$
$$\neg p \lor q \quad \lor \quad \neg q \lor \neg p$$
$$(p \lor \neg p) \quad \lor \quad (q \lor \neg q)$$
$$true$$

This goes against the common sense idea that two variables can be independent.

9.3 MODAL LOGIC

One can extend a logical system by adding modal operators to the set of allowable operators, including the ideas of necessity and possibility, in addition to quantification. These may be useful when describing information needs or documents; for example, whether a need or document necessarily addresses a specific topic, or whether it might be of use or a user might find useful a certain topic. Modal logics incorporate modal operators, including necessity and possibility. Statements with the modal necessity operator are *apodeictic*, necessarily true and clearly demonstrable, while statements with the possibility operator are *problematic*, uncertain or possibly true.

The notation and systems described here are those based on *Survey of Symbolic Logic* by C. I. Lewis (1918), and later expanded in *Symbolic Logic* (1932), the first modern exploration of the subject and usually considered as the foundation for the modern study of modal logic. The possibility of x is understood here as representing that in some situation or *world*, x is true. The necessity of x implies that in all situations or worlds x is true. Stating that x is necessary may be written either as $\Box x$ or Lx, while a statement that x is possible may be represented by $\Diamond x$ or Mx, where M stands for *möglich* (Hughes & Cresswell, 1968). Using Hx to again represent that x is human and Fx to represent that x is female, it is the case that if x is human then x might be female, $Hx \supset \Diamond Fx$, and if an x is human it must be either female or not female, $\Box(Hx \supset (Fx \lor \neg Fx))$.

Modal logic systems may be built upon propositional logic and either the necessity or possibility operators. Possibility may be defined as "it is not necessary that it is not....," that is, $\Diamond p \equiv \neg \Box \neg p$. Similarly, necessity may be defined as "it is not possible that it is not.....," or $\Box p \equiv \neg \Diamond \neg p$. For these reasons, most commonly examined modal logic systems accept the following:

$$\neg \Box p \quad \equiv \quad \Diamond \neg p$$
$$\neg \Diamond p \quad \equiv \quad \Box \neg p.$$

They also accept the notion that if $\Diamond p$, then in some world, somewhere, p is true (Gensler, 1990).

Strict Implication

Strict implication, denoted as $p \rightarrow q$, should be understood as p strictly implies q but not that p materially implies q. Strict implication states that it is *impossible* that the first statement is true and the second statement false. Thus, for material implication,

$$p \supset q \equiv \neg(p \wedge \neg q),$$

while for strict implication

$$p \rightarrow q \equiv \neg\Diamond(p \wedge \neg q).$$

Implication may be represented probabilistically so that $a \rightarrow b$ occurs with the probability $\Pr(b|a)$. By using different means of estimating these probabilities (Crestani & van Rijsbergen, 1995), one can provide quantitative rankings of entities in logical systems based upon modal logics.

Problems with Strict Implication

Sanford (1989) suggests the following examples of problems extant with strict implication. The statement

$$\neg\Diamond p \rightarrow (p \rightarrow q)$$

implies that if something is not possible, then one can conclude anything. Furthermore, a tautologically false statement, such as $p \wedge \neg p$, allows us to conclude anything:

$$(p \wedge \neg p) \rightarrow q.$$

If something is necessary, it is strictly implied by anything,

$$\Box p \rightarrow (q \rightarrow p),$$

while a tautologically true statement is strictly implied by anything:

$$p \rightarrow (q \vee \neg q).$$

Consistency and Contingency

Using the two proposed modalities, it becomes possible to represent other operations or modalities than just necessity and possibility. For example, one may represent "it is impossible that p" by either $\neg\Diamond p$ or $\Box\neg p$. Consistency may be represented, with "p is consistent with q" being translated into $\Diamond(p \wedge q)$. Inconsistency is similarly represented as $\neg\Diamond(p \wedge q)$. Representing consistency is useful where one wishes to show that particular terms or concepts may occur together or that they may not co-occur.

Contingencies may also be represented using modalities. A contingent statement, p, may be expressed as $\Diamond p \wedge \Diamond \neg p$, that it is possible that statement p is true and it is possible that statement p is not true. A statement that is true might not always be true or be true in all worlds. This may be expressed as $p \wedge \Diamond \neg p$, which claims that p is true and it is possible that at some point in time, or in some world, it is possible that p is false. Contingencies may be used to indicate whether a proposition is true in some cases, such as "Caitlyn's blouse is purple," or whether it is not contingent, such as "X is X."

Principles of Reduction

Using the Lewis definition of strict implication discussed earlier, and if we use $\Box$ as defined as or syntactically equivalent to $\neg \Diamond \neg$, then the *weak reduction principle* can be expressed as either of the following:

$$\neg \Diamond \neg p \rightarrow \neg \Diamond \neg \neg \Diamond \neg p$$
$$\Box p \rightarrow \Box \Box p.$$

The weak reduction principle excludes from a logical system the possibility that if something is necessary that it is contingently or only possibly necessary; the principles dictates that it is necessarily necessary.

The *strong reduction principle* may be expressed in any of the following three ways:

$$\Diamond p \quad \rightarrow \quad \neg \Diamond \neg \Diamond p$$
$$\Diamond p \quad \rightarrow \quad \Box \Diamond p$$
$$\neg \Box p \quad \supset \quad \Box \neg \Box p \quad \equiv \quad \Diamond \neg p \supset \Box \Diamond \neg p.$$

Adding this principle in addition to its weaker counterpart places the requirement upon a logical system that any proposition be necessarily true or false. Stating that $\Diamond p$ strictly implies $\Box \Diamond p$ forces all possibilities to be necessarily possible, and, combined with $\Box p \rightarrow \Box \Box p$ from the weak reduction principle forces all modal propositions to be necessarily true or false.

Accepting both the strong and weak principles of reduction allows all modalities to be collapsed from nth order modalities to first order modalities in which only a single modal symbol is present. The eight modal expressions using expression p are $\Box p$, $\neg \Box p$, $\Diamond p$, and $\neg \Diamond p$, as well as similar modalities applied to $\neg p$.

Systems of Modal Logic

There have been numerous logical systems developed containing modal operators. C. I. Lewis developed a set of systems S1 through S5 such that all theorems in S1 are in S2, all theorems in S2 are in S3, and so forth. Thus, S5 can be understood as containing the largest set of theorems and thus be the most complex.

The systems may best be understood by examining the statement added to each system, with a conventional propositional logic existing below S1 (Sanford, 1989):

$$(S5) \qquad \Diamond p \quad \supset \quad \Box \Diamond p$$
$$(S4) \qquad \Box p \quad \supset \quad \Box \Box p$$
$$(T) \quad \Box((p \to q) \quad \to \quad (\Box p \supset \Box q))$$
$$(S3) \qquad (p \to q) \quad \to \quad (\Box p \Box q)$$
$$(S2) \qquad \Diamond(p \wedge q) \quad \to \quad \Diamond p$$
$$(S1) \qquad (p \to q) \quad \supset \quad (\Box p \supset \Box q).$$

S3 and T are not hierarchically related; they should both be seen as different paths from S2 to S4. System T, $S4$, and $S5$ allow $\Box p \to p$ in a given world. Only system $S5$ allows $\Box p$ to a p in another world (Gensler, 1990).

T

Robert Feys (1965) called a particular modal system "T." It uses strict implication and possibility and can thus leave "necessity" undefined (Prior, 1967).

In addition to the rules of propositional logic, it contains

$$\Diamond p \equiv \neg \Box \neg p$$
$$p \to q \equiv \Box(p \supset q)$$
$$\Box p \supset p$$
$$\Box(p \supset q) \supset (\Box p \supset \Box q).$$

In addition, the rule of "necessitation" is used: $A \to \Box A$ (Turner, 1984; Prior, 1967).

If the fundamental rule for T is viewed as $\Box p \to p$, then one can intuitively accept the notion that if it is necessarily true that p, then p should be *true*. If one treated the $\Box$ operator as "believes that," it becomes difficult to accept this fundamental axiom of T.

S4

The system $S4$ may be derived from T by adding $LA \to LLA \equiv \Box A \to \Box \Box A$ to T (Turner, 1984; Prior, 1967). Thus, what is necessary is necessarily so (Turner, 1990). One may add either $\Diamond \Diamond p \to \Diamond p$ or $\Box p \to \Box \Box p$ to more basic systems (such as $S3$) to produce $S4$. Yonemitsu (Feys, 1965) suggests that $S4$ can be produced from $S3$ through the addition of $\Box(p \to p)$. The system may be understood as the lower level systems with the weak reduction principle added.

S4 has been suggested by Gödel to be equivalent to a basic propositional logic with the following axioms added:

$$\Box p \supset p$$
$$\Box p \supset (\Box(p \supset q) \supset \Box q)$$
$$\Box p \supset \Box\Box p.$$

This also requires a rule stating that $p \supset \Box p$.

The following modalities are acceptable in $S4$: p, $\Box p$, and $\Diamond p$, as well as their negations, as in S5, and $\Box\Diamond p$, $\Diamond\Box p$ $\Diamond\Box\Diamond p$, and $\Box\Diamond\Box p$, as well as their negatives (and the expressions applied to $\neg p$).

All the deductions one can make in S4 follow from the premises in S5. In fact, all the theorems in one of Lewis' set of systems S5, S4, S3, S2, and S1 can be deduced from any higher numbered system. The following four reduction laws exist in S4: :

$$\Diamond p \overset{\Box}{\equiv} \Diamond\Diamond p$$
$$\Box p \overset{\Box}{\equiv} \Box\Box p$$
$$\Diamond\Box p \equiv \Diamond\Box\Diamond\Box p$$
$$\Box\Diamond p \equiv \Box\Diamond\Box\Diamond p.$$

These are somewhat more restrictive reduction laws than those found in S5, which allows far more reduction. Consistent with S4, one can maintain that "all necessary propositions are necessarily necessary" in S4 without maintaining that "all possible propositions are necessarily possible," as is maintained in S5 (Hughes & Cresswell, 1968).

$S4$ can prove useful if the $\Box$ operator is treated as "knows that," since if one knows something, one knows that one knows something, as with $\Box p \rightarrow \Box\Box p$.

S5

S5 is a modal system consistent with the basic theorems of $S4$ and the strong reduction principle. S5 can thus be derived by the addition of the following axiom, representing the strong reduction principle:

$$\neg\Box p \supset \Box\neg\Box p.$$

One can similarly add $\Diamond A \rightarrow \Box\Diamond A$ to S4 to get S5 (Turner, 1984; Feys, 1965) Unlike lower level and more lenient modal systems, $S5$ allows that if it is necessary that p in any world then it is necessary that p is true in all worlds (Gensler, 1990). $S5$ is an economical system, allowing a relatively small number of modalities. The following modalities are acceptable: p, $\Box p$, and $\Diamond p$, as well as their negations. All strings of modalities can be simplified by deleting modalities to the left. Thus, $\Box\Diamond\Box p$ becomes simply $\Box p$. Reduction may use one of the following four reduction laws that exist in S5 (Hughes & Cresswell,

1968):

$$\Diamond p \; \overset{\Box}{\equiv} \; \Box\Diamond p \overset{\Box}{\equiv} \Diamond\Box\Diamond p$$

$$\Diamond p \; \equiv \; \Diamond\Diamond p$$

$$\Box p \; \overset{\Box}{\equiv} \; \Diamond\Box p \overset{\Box}{\equiv} \Box\Diamond\Box p$$

$$\Box p \; \equiv \; \Box\Box p.$$

Thus, whatever is possible is necessarily possible. If p is necessary in all worlds or situations, the one can conclude that p is true. One can conclude that "all worlds have the same necessary truths" (Gensler, 1990).

Combining Quantification and Modalities

In some instances it becomes necessary (or desirable) to combine modal logic with quantifications from predicate logic. Consider a statement such as "Everyone has a chance of winning" which may be translated as "For all x, possibly x will win." This is *de res*, as the scope of the modality is the set of objects (*res*). Given the similar sentence, "There is a chance that everyone will win," one arrives at a translation such as "Possibly for all x, x will win." This is a *de dictum* modality, because the scope of the modality is the proposition (*dictum*), as opposed to an object or objects.

There are a number of relationships that can hold between modalities and quantification:

$$\forall x \Box Px \longleftrightarrow \Box \forall x Px \tag{9.3}$$

$$\exists x \Box Px \longrightarrow \Box \exists x Px \tag{9.4}$$

$$\forall x \Diamond Px \longleftarrow \Diamond \forall x Px \tag{9.5}$$

$$\exists x \Diamond Px \longleftrightarrow \Diamond \exists x Px. \tag{9.6}$$

Moving down the two columns of logical expressions, one can find that the expressions in Equation 9.3 imply both the expressions in Equations 9.4 and 9.5 in their respective columns. The expressions in Equations 9.4 and 9.5 both separately entail the expressions in the same column in Equation 9.6. The arrows between the columns represent entailments or if and only if relationships that exist between the expressions. The expressions in the right column are easily deducible–the ones on the left are problematic (Kneale & Kneale, 1984).

9.4   TEMPORAL LOGICS

Temporal logics are useful at representing the need for information that might or will have certain characteristics, or had or might have had a characteristic.

We may use the following operators to represent some temporal aspects of logic (Turner, 1984):

$\exists_F p$ there exists a future time at which p is true
$\forall_F p$ for all future times p will be true
$\exists_P p$ there exists a past time at which p was true
$\forall_P p$ for all past times p was true.

The following modal relationships are useful:

$$\forall_F p \;\equiv\; \neg\exists_F\neg p$$
$$\forall_P p \;\equiv\; \neg\exists_P\neg p.$$

A minimal temporal logic may be used as a basis upon which to build further logics, and may be useful in and of itself. Turner begins such a logic with the following axioms:

$$p, \qquad \text{where } p \text{ is a tautology} \qquad (9.7)$$
$$\forall_F(p \supset q) \;\supset\; (\forall_F p \supset \forall_F q) \qquad (9.8)$$
$$\forall_P(p \supset q) \;\supset\; (\forall_P p \supset \forall_P q) \qquad (9.9)$$
$$p \;\supset\; \forall_P \exists_F p \qquad (9.10)$$
$$p \;\supset\; \forall_F \exists_P p \qquad (9.11)$$
$$\forall_F p, \qquad \text{if } p \text{ is an axiom} \qquad (9.12)$$
$$\forall_P p, \qquad \text{if } p \text{ is an axiom.} \qquad (9.13)$$

This system requires traditional propositional logic and modus ponens. Equations 9.8 and 9.9 above state that for all futures (or pasts), such that p implies q, all future (past) ps imply all future (past) qs. Equations 9.10 and 9.11 state that p implies that for all pasts (futures), there is a future (past) where p holds. Equations 9.12 and 9.13 state that if p is a valid axiom, then "in all futures p" and "in all pasts p" are valid axioms

Several additions to a minimal temporal logic will produce more useful semantics.

Transitivity

Transitivity implies if at some point in the future there is a point in the future such that p implies that at some point in the future p. More formally, $\exists_F \exists_F p \supset \exists_F p$. Less formally, if it is 9 A.M., there is a future time (e.g., 10 A.M.), such that there is a time in the future beyond 10 A.M. (e.g., 11 A.M.) implies that there is a time beyond 9 A.M. that is in the future (e.g., 11 A.M.).

Linearity

The principle of *backwards linearity* simply states that if there are two points in the past, either they are simultaneous or that one precedes the other,

$$(\exists_P p \wedge \exists_P q) \supset [\exists_P(p \wedge q) \vee (\exists_P(p \wedge \exists_P q) \vee \exists_P(\exists_P p \wedge q))].$$

Here $\exists_P(p \wedge q)$ denotes a simultaneous action at a particular time, $\exists_P(p \wedge \exists_P q)$ denotes action q being true before p and $\exists_P(\exists_P p \wedge q)$ denotes action p being true before q is true.

The principle of *forward linearity* similarly states that either two propositions p and q occur simultaneously in the future or one true proposition precedes the other;

$$(\exists_F p \wedge \exists_F q) \supset [\exists_F(p \wedge q) \vee (\exists_F(p \wedge \exists_F q) \vee \exists_F(\exists_F p \wedge q))].$$

One may wish to represent that fact that time has no beginning and no end as

$$\forall_F p \quad \supset \quad \exists_F p$$
$$\forall_P p \quad \supset \quad \exists_P p.$$

One may also wish to represent that time may be represented as a linear continuum,

$$\exists_F p \quad \supset \quad \exists_F \exists_F p$$
$$\square(\forall_F p \supset \exists_P \forall_F p) \quad \supset \quad (\forall_F p \supset \forall_P p).$$

Temporal logic is most beneficial when what we refer to as "documents" are dynamic entities. For example, Internet web pages are increasingly created at the moment a query is received. One can describe either the past or future of these changing objects. Dynamic entities may be processes that produce output that changes over time, based on either changing input or the intrinsic complexity of the process. These changing entities may also be human information sources. Temporal logic may also be useful within documents themselves, stating that a predicate was true or will be true at some point in the past or the future.

9.5 LOGIC AND PROBABILITY

Individual events have probabilities of occurrences. Statements describing events, such as logical propositions making statements about the world, have associated probabilities that represent the degree to which the state of nature described by the proposition holds. While individual, atomic propositions have clearly defined interpretations, the meanings and values of probabilities associated with logical operators is not as obvious.

A logical proposition can be described probabilistically over the universe of logical values associated with the proposition. It is important that the probabilities described here are understood as the probability of a given proposition being true, taken from the universe of propositions from which the probabilities are being developed. The probability associated with proposition p is $\Pr(p)$ and the probability associated with the negation of p is $1 - \Pr(p)$. The relationships between logical expressions and their probabilities are provided by the following:

Definition 9.1 (Probability of a logical proposition) *The probability of proposition p is denoted as* $\Pr(p)$.

Theorem 9.1 (Probability of a negated proposition) *The probability of proposition $\neg p$ is* $1 - \Pr(p)$.

Theorem 9.2 (Probability of conjoined propositions) *The probability of a set of conjoined propositions $p_1 \wedge p_2 \wedge \cdots \wedge p_{n-1} \wedge p_n$ is*

$$\Pr(p_1, p_2, \ldots, p_{n-1}, p_n) \tag{9.14}$$

Theorem 9.3 (Probability of disjoined propositions) *The probability of a set of disjoined propositions $p_1 \vee p_2 \vee \cdots \vee p_{n-1} \vee p_n = \neg(\neg p_1 \wedge \neg p_2 \wedge \cdots \wedge \neg p_{n-1} \wedge \neg p_n)$ is*

$$1 - \Pr(-p_1, -p_2, \ldots, -p_{n-1}, -p_n) \tag{9.15}$$

where $-p$ represents the values not in p.

Theorem 9.4 (Probability of logical implication) *The probability of the proposition $(p \supset q) \equiv (q \vee \neg p)$ is* $1 - \Pr(\neg q, p)$.

Any Boolean expression or any expression in propositional logic can be stated in Conjunctive Normal Form (CNF). If any logical expression in CNF can be expressed probabilistically, then the probability of any Boolean expression or logical rule can be computed and used to determine the expected retrieval performance given a query. The probability is computed as the joint distribution of the sets or probabilities of each of the conjoined elements. The disjunction in each of the conjunctions may be computed directly.

Consider the query $A \wedge (B \vee C)$ in CNF. We find that $B \vee C = \neg(\neg B \wedge \neg C)$. The probability associated with the full expression is the joint probability

$$\Pr\big(A, \neg(\neg B \wedge \neg C)\big).$$

If disjoined features in the conjunctive elements are statistically independent, this probability becomes

$$\Pr(A)\,(1 - \Pr(\neg B, \neg C)).$$

Here, $\Pr\big(\neg B, \neg C\big)$ is the joint probability that neither B nor C occurs.

Theorem 9.5 (Probability of a CNF expression) *Consider an expression in CNF:*

$$(e_{11} \vee e_{12} \vee \ldots \vee e_{1a}) \wedge (e_{21} \vee e_{22} \vee \ldots \vee e_{2b})$$
$$\wedge \ldots \wedge (e_{m1} \vee e_{m2} \vee \ldots \vee e_{mn}) \tag{9.16}$$

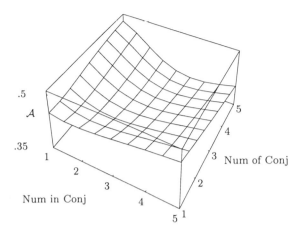

.5

$\mathcal{A}$

.35

Num in Conj

1

2

3

4

5

1

2

3

4

5

1

2

3

4

5

Num of Conj

Figure 9.1.   The $\mathcal{A}$ performance as the number of items in each conjunction and the number of conjunctions varies. For each item in the disjunction, $p = .5$ and $t = .4$, and the probabilities for the both relevant and all documents are computed consistent with the Boolean expression.

where there are m conjunctive elements and the number of disjunctive elements is a for the first conjunctive element, b for the second conjunctive element, and n for the last conjunctive element. Computing the probability for Equation 9.16 requires the full evaluation of the set relations in the equation. If each element in a disjunction has the same probability and the conjoined components are statistically independent, then the probability of the entire statement is simplified to:

$$(1 - \Pr(-e_{11})^a)(1 - \Pr(-e_{21})^b) \cdots (1 - \Pr(-e_{m1})^n)$$

where $\Pr(-e_i)^j$ is the probability $\Pr(-e_i)$ taken to the jth power.

The performance using this estimate is shown in Figure 9.1. We assume in the Figure that each disjunction is the same size and the size of each disjoined expression varies, as does the number of conjoined disjunctions. The formula used here is thus

$$\left(1 - \Pr(-e)^{\text{num in conjunction}}\right)^{\text{num of conjunction}}.$$

Equal Performance for Boolean and Probabilistic Ranking

The ranking of documents resulting from a Boolean query such as x **and** y may be emulated probabilistically with a probabilistic ranking of documents similar to what is obtained with the query for terms x and y, assuming term dependence (Losee, 1997b). To obtain a probabilistic ranking that is equivalent to the Boolean ranking, it is necessary that the probabilistic ranking be able to limit the ranking to two categories of documents as does the Boolean ranking. In the case of the Boolean **and**, the probabilistic model must place documents with both terms in one group and documents with only one or neither of the terms in a second group, with all in the second group having the same document weight. We may do this consistent with the Bahadur-Lazarsfeld Expansion (Equation 7.4), noting that the degree of dependence may be modified, assuming a set of probabilities, by modifying the expected product of the two random variables, denoted here as $c = E(ab)$.

Theorem 9.6 (Break-even expected joint probability for "and") *If we assume that $c = E(ab)$ and c_t is computed from the entire database, the value c_p^{and}, the break-even c value for relevant documents at which probabilistic and Boolean rankings are the same, when 2 terms in the query are conjoined in the Boolean query, is computed as*

$$c_p^{\text{and}} = 1 + \frac{-1 + c_t + p - c_t p}{1 - t}, \tag{9.17}$$

if we assume that the p and t values for both terms are equivalent.

To achieve this ranking, it is necessary that

$$\Pr(x = 1, y = 0) = \Pr(x = 0, y = 1)$$
$$= \Pr(x = 0, y = 0) \neq \Pr(x = 1, y = 1).$$

This may be obtained if we equate the probabilities of terms $x = 1$ and $y = 1$, making $\Pr(x = 1, y = 0) = \Pr(x = 0, y = 1)$. Next, dependence between terms must be determined such that $\Pr(x = 1, y = 0) = \Pr(x = 0, y = 0) = \Pr(x = 0, y = 1)$. To do this, the c value for the Boolean **and**, c^{and}, must be set so that these constraints are met, where c in this section denotes the expected proportion of documents with both terms, thus $c = E(ab)$.

When one ranks documents assuming term dependence, the true values of p, t, and c_t are used, and c_p is estimated as in Equation 9.17, giving the same ranking as would be obtained with a Boolean query using the **and** operator.

EXERCISE 9.2. [*Moderate*] What would be the formula for c_p^{or}?

EXERCISE 9.3. [*Moderate*] Consider the following 100 document profiles and frequencies:

Profile	Num. in All Docs.	Num. in Rel. Docs.
1 1	30	25
1 0	20	5
0 1	20	4
0 0	30	5

Given the Boolean query X **and** Y, where X is the left feature in the profile and Y the feature on the right, what is the $\mathcal{A}$ value? For the Boolean query X **or** Y, what is the $\mathcal{A}$ value?

9.6   FILTERING

Boolean expressions can serve as filters for both documents represented as lists of features and for documents with their contents themselves expressed as Boolean expressions. In additional, multiple filters may be applied successively to a single set of documents, with those documents meting the requirements of all filters being retrieved. Assume $X = b_1 \wedge b_3$ and $Y = b_1$ and there exist 3 Boolean reduced atomic features b_1, b_2, and b_3.

The set of Boolean statements that can be consistent with X are

$$b_1 \quad \wedge \quad b_2 \quad \wedge \quad b_3$$
$$b_1 \quad \wedge \quad \neg b_2 \quad \wedge \quad b_3.$$

These can be combined to yield, in disjunctive normal form,

$$X = (b_1 \wedge b_2 \wedge b_3) \vee (b_1 \wedge \neg b_2 \wedge b_3).$$

A larger set of statements are consistent with Y:

$$b_1 \quad \wedge \quad b_2 \quad \wedge \quad b_3$$
$$b_1 \quad \wedge \quad \neg b_2 \quad \wedge \quad b_3$$
$$b_1 \quad \wedge \quad b_2 \quad \wedge \quad \neg b_3$$
$$b_1 \quad \wedge \quad \neg b_2 \quad \wedge \quad \neg b_3.$$

These can be combined to yield

$$Y = (b_1 \wedge b_2 \wedge b_3) \vee (b_1 \wedge \neg b_2 \wedge b_3) \vee (b_1 \wedge b_2 \wedge \neg b_3) \vee (b_1 \wedge \neg b_2 \wedge \neg b_3).$$

The features common to these two sets X and Y are those included in X,

$$b_1 \quad \wedge \quad b_2 \quad \wedge \quad b_3$$
$$b_1 \quad \wedge \quad \neg b_2 \quad \wedge \quad b_3.$$

Documents with these features would be retrieved or passed through, given the two filters X and Y.

The degree of similarity here may be estimated by the number of expressions held in common or that are consistent with the compared items.

Predicate Logic Filtering

The actions that take place in filtering may be described using predicate logic. A query is understood as a statement of the set of characteristics or nature of documents to be passed by a filter. A document is evaluated based upon the characteristics that it holds, and the decision is made to pass a document through the filter or to "filter it out." The filter determines what characteristics are examined in the objects that pass through it; thus, the filter may be said to imply the characteristics of the object passed.

An information filter may be described using formal logic by noting which predicates (and the actions or states of nature they are based on) are consistent in both the query and the document. At the core of a simple filter, with a query providing *all* and *only* the characteristics of documents to be passed, a document should be passed if and only if the query and the document both have the characteristic or both lack the characteristic. More formally,

$$\forall q \ \forall d \big(\forall c(Qqc \wedge Ddc) \equiv Pqd\big), \tag{9.18}$$

where d denotes documents, q denotes queries, and c denotes characteristics representing aspects of queries or documents such as terms. In addition, characteristics are assumed to be expressed in one way; that is, by c which is assumed to be constant for a particular query. The Qqc predicate is true when the characteristics are c in query q which a document passed by the filter must contain. Predicate Ddc is true when document d that is passed through the filter has characteristic c, and Pqd is true or false depending on whether document d is passed through by the filtering mechanism consistent with query q. This is similar to the mechanism proposed by Watters (1989). Following convention, documents and query predicates are treated as being distinct, although the author believes that for many purposes, they may both be treated as "about" predicates.

This model assumes that the query describes all and only the characteristics of those documents to be passed by the filter, a perfect match between query and document. More realistic is a situation where a query completely specifies some aspect of a document, requiring that for a document to be passed through the filter, it must contain certain features and must not contain certain other features. However, the querier may not care, for example, whether a book has a red or a blue cover; this characteristic is irrelevant to the querier's interests. But they may care whether it addresses the topic *information retrieval*.

Boolean expressions may serve as filters, allowing control to be exercised over which documents pass through the filter. A filtering Boolean expression may be placed into either conjunctive normal form or disjunctive normal form. Because each leads to a different conceptual approach to filtering, both are examined.

Beginning with an expression in conjunctive normal form, a logical expression describing the filtering action is

$$\forall q \ \forall d(\forall \hat{c}(Qq\hat{c} \supset Dd\hat{c}) \equiv Pqd),$$

which may be rewritten as

$$\forall q \ \forall d(\forall \hat{c}(\neg Qq\hat{c} \vee Dd\hat{c}) \equiv Pqd), \tag{9.19}$$

Table 9.1. The query *apples* **and** *oranges* in CNF (with CNF filter Equation 9.19) and a set of documents. Truth values being multiplied are $(\neg Qqc \vee Ddc)$.

Source	apples	oranges	dogs	cats	P_{qd}
Query	Y	Y	N	N	
D_1	Y	Y	N	Y	$T \times T \times T \times T = T$
D_2	Y	Y	Y	Y	$T \times T \times T \times T = T$
D_3	N	Y	Y	N	$F \times T \times T \times T = F$

where the query q is a conjunction of disjoined characteristics, each conjoined sub-expression being denoted as a characteristic $\hat{c}$. To pass the document, each conjoined sub-expression is in the document or it is not in the query.

A query expressed in disjunctive normal form, a disjunction of conjunctions, may be similarly incorporated into a logical rule describing the filter's actions:

$$\forall q \ \forall d (\exists \check{c} (Qq\check{c} \wedge Dd\check{c}) \equiv Pqd), \tag{9.20}$$

where the query q is a disjunction of conjoined characteristics, each disjoined sub-expression being denoted as a characteristic $\check{c}$

Example. Consider the situation where a query contains the terms *apples* and *oranges* and a document d_1 has the terms *apples, oranges, dogs*, and *cats* (Table 9.1). The right side of the table shows whether the document is passed. When a CNF query is used with Equation 9.19, the combination of the conjoined values may be multiplied, as in the table.

Synonyms and Related Terms

Synonyms for terms can be used in logical formulae placed into CNF, expanding the earlier filtering model. Consider

$$\forall q \ \forall d (\forall \hat{c} ((Qq\hat{c} \vee QRq\hat{c}) \supset (Dd\hat{c} \vee DRd\hat{c})) \equiv Pqd), \tag{9.21}$$

where QR and DR are *related term* predicates similar to the Q and D predicates, relating terms produced by other logical operations that are assumed to be conjoined with filter description statements such as Equation 9.21.

Modalities

Consider the following rule for passing documents:

$$\forall q \ \forall d \Big((\forall c (Qqc \supset Ddc) \wedge \forall c' \Diamond (Qqc' \supset Ddc')$$
$$\wedge \ \forall c'' \Box (Qqc'' \supset Ddc'')) \equiv Pqd \Big) \tag{9.22}$$

where d denotes documents, q denotes queries, and c, c', and c'' each denote characteristics in a set representing aspects of queries or documents. Those characteristics which are in both query and document are denoted by c, those characteristics that are *possibly* in both query and document are in the set c', and those characteristics that are *necessarily* in both query and document are in the set c''. The Qqc predicate is true when the characteristics of query q are the same characteristics as c which a document passed by the filter must contain.

9.7 LOGIC AND QUANTITIES

Logic is usually viewed as being based upon truth and falsehood. While this condition may be relaxed with multivalued logics, one may use traditional two valued logics in situations addressing quantitative considerations.

A probabilistic or vector model of a filter's actions may be based upon a similarity measure of the conceptual distance between the query and the document. In addition, a threshold value for characteristics c must be exceeded by the similarity measure if the document is to be passed by the filter. Consider

$$\forall q \; \forall d \Big(G\big(\forall c \; S(Qqc, Ddc), Tq\big) \equiv Pqd \Big) \qquad (9.23)$$

where $S(q, d)$ is the similarity of q and d, Tq is the threshold function, and $G(x, y)$ is true if x is greater than y.

Logical expressions may be similar in different ways. Two expressions may have the same meaning. In a more pragmatic sense, two expressions may have the same result, or, on a more formal level, two expressions may be similar or identical if they are very similar or identical when placed into a particular normal form, such as CNF.

The simple match function is useful as a tool for counting the similar features between, for example, a query and a document. The value of the simple match function is inherently non-binary, that is, it of necessity returns a value that ranges from 0 to 1. When a single characteristic is considered, the simple match may be modeled logically. For a term c, one may use

$$(Qqc \wedge Ddc) \vee (\neg Qqc \wedge \neg Ddc).$$

This can produce one of two values, *true* or *false*. These correspond to two values that may be obtained from the simple match function, 1 and 0. If there were two characteristics, however, the simple match function could produce one of three values: 1, 1/2, or 0.

Matching on at Least m Cases

Matching may be based on the number of predicates that are true. Matching is said to occur when at least 2 predicates f are true, then

$$\exists x \exists y (f_x \wedge f_y \wedge (x \neq y)).$$

This can be generalized to note that at least n predicates f are true when

$$\exists x_1 \exists x_2 \ldots \exists x_n \Big(f_{x_1} \wedge f_{x_2} \wedge \ldots \wedge f_{x_n} \wedge$$
$$(x_1 \neq x_2) \wedge (x_1 \neq x_3) \wedge \ldots \wedge (x_1 \neq x_n) \wedge$$
$$(x_2 \neq x_3) \wedge (x_2 \neq x_4) \wedge \ldots (x_{n-1} \neq x_n) \Big).$$

This may be simplified further to

$$L_{x,n} = \exists x_1 \exists x_2 \ldots \exists x_n \Big(\prod_{i=1}^{n} f_{x_i} (\prod_{i=1}^{n-1} \prod_{j=2}^{n} x_i \neq x_j) \Big).$$

Matching on no More than m Cases

In other cases, it is desirable to determine whether there are at most n predicates f that are true. If there are more, they must have been counted twice.

$$M_{x,n} = \forall x_1 \forall x_2 \ldots \forall x_{n+1} \Big(\big(f_{x_1} \wedge f_{x_2} \wedge \ldots \wedge f_{x_{n+1}} \big) \supset$$
$$\big((x_1 = x_2) \vee (x_1 = x_3) \vee \ldots \vee (x_1 = x_{n+1}) \vee (x_2 = x_3) \vee \ldots \vee$$
$$(x_2 = x_{n+1} \vee \ldots \vee (x_n = x_{n+1})) \big) \Big). \quad (9.24)$$

This may be interpreted as stating that if there are more than n matching characteristics, then two of these characteristics must in fact be the same.

Matching on Exactly m Cases

Using these predicates that determine whether there are at least n true predicates or at most n true predicates, one can determine whether an expression has exactly n true predicates:

$$S_{x,n} = M_{x,n} \wedge L_{x,n} \quad (9.25)$$

Using this allows one to determine whether a document is similar to a query to a certain degree, the core need of filtering and retrieval.

9.8   RELATIONS AND FACT RETRIEVAL

One way of describing factual knowledge begins with the set of values that a variable may take. For example, a variable *academic discipline* might have a domain of 40 or 50 possible values, including *physics, botany, Latin,* etc. We assume here that the variables can take on exactly one value at any given time.

An ordered set of related variables is referred to as a *relation*. A set of values for the relationship, that is, a set of specific values, is referred to as a *tuple*, with an n variable tuple being referred to as an n-tuple. A relation may be viewed graphically as a two dimensional table, with each column representing a specific variable (called an *attribute*) and each row a specific tuple. One or more of the variables are thought of as the *key* and serve as the access point to the relation by allowing one to access one or more specific tuples.

A relation is normalized through any of a number of processes that decrease the functional and statistical relationships between non-key variables, except for the dependence induced by the presence of the key variable itself (Lee, 1987). For example, Table 9.2 contains name, address, and house color information in an unnormalized form. There is a measure of redundancy in Table 9.2,

Table 9.2. The first table contains unnormalized sample data, while the second and third tables contain the same data normalized.

Name	Address	House Color
Bob	106 Shady Lane	green
Lee Ann	106 Shady Lane	green
Caitlyn	106 Shady Lane	green
Ben	108 Shady Lane	yellow
Elizabeth	108 Shady Lane	yellow

as several tuples in the table show that 106 Shady Lane is green and several show that 108 Shady Lane is yellow. By *normalizing* the data, one arrives at Tables 9.3a and 9.3b:

Table 9.3a. Name and Address in normalized table.

Name	Address
Bob	106 Shady Lane
Lee Ann	106 Shady Lane
Caitlyn	106 Shady Lane
Ben	108 Shady Lane
Elizabeth	108 Shady Lane

Table 9.3b. Address and House Color in normalized table.

Address	House Color
106 Shady Lane	green
108 Shady Lane	yellow

which do not have this redundancy. Normalized tables are *joined* together when needed, bringing together the needed facts.

Probabilities in a Relation

The probability of a particular instance of a relation r_i from domain D of instances can be computed as $\Pr(r_i) = ||r_i||/||D||$, where the cardinality notation $||x||$ represents the number of elements in x. Given this probability, and our earlier definitions for p, t, $\mathcal{A}$, and $\mathcal{Q}$, the ASL may be computed. While this may appear somewhat trivial for simple relations, it becomes more complex (and more interesting) when relations are joined together and the combined relation is retrieved.

The retrieval of a relevant record with k will have the probabilities $p = \Pr(k|rel)$ and $t = \Pr(k)$. Using Equation 8.2, one can compute $\mathcal{A}$. Thus, if there are 100 records, 10 of them with the key and 1 with the attribute of interest (and thus relevant), then $p = 1$ and $t = .1$. One may thus compute $\mathcal{A} = (1 - 1 + .1)/2 = .05$.

If there are two normalized relations, and we wish to join them, such as would be the case with the normalized relations in the Table and a query such as *What is the color of Caitlyn's house?*, we are essentially searching through the unnormalized table, and the $ASL = 2$.

EXERCISE 9.4. [*Easy*] If one retrieved the persons living at 106 Shady Lane from Table 9.2, and Lee Ann's entry was the only one that was relevant, what would be the ASL?

9.9 THREE-VALUED LOGICS

Logic has traditionally based itself on truth and falsehood, or two other opposing values. These systems may be expanded to be consistent with a multi-valued world (Turner, 1984; Kneale & Kneale, 1984).

Three-valued logics usually contain both values representing *true* and *false*. The systems often differ in how a third, sometimes "middle" value is represented. This allows additional precision in the expression of the information need and should result in superior filtering and retrieval performance.

Kleene's Strong Truth Valued Logic

Kleene proposed a three-valued logic with the following values:

> T *true*
> F *false*
> U *undecided* or *undefined.*

A proposition has a value *true* or *false* only when one has certain knowledge. Undecided represents an intermediate truth value that occurs when one doesn't have certain knowledge. The differences between three-valued logic systems produce different rules for the logical operators, based on the differing interpretations for this intermediate value.

Kleene's system has the following truth table:

p	q	$\neg p$	$p \wedge q$	$p \vee q$	$p \equiv q$	$p \supset q$
T	T	F	T	T	T	T
T	F	F	F	T	F	F
F	T	T	F	T	F	T
F	F	T	F	F	T	T
T	U	F	U	T	U	U
F	U	T	F	U	U	T
U	T	U	U	T	U	T
U	F	U	F	U	U	U
U	U	U	U	U	U	U

Probabilities

The probability that a statement is *true* may be computed under two sets of assumptions. The first is that statements with undefined truth values are not to be included in the computation of the probability. In this case, the probability of a proposition is the proportion of *true* statements from among those statements with the truth values of either *true* or *false*. This can serve as the upper bounds for the second probability, that a statement is *true* from among all propositions, those that are *true* or *false* as well as those statements whose truth values are undecided. When probabilities may be computed by including or excluding the third values, the A value may be computed using techniques described in preceding chapters.

Lukasiewicz's Logic

Lukasiewicz's logic provides three values: truth, falsehood, and indeterminacy. I denotes that a statement does not have a logical value, not that we don't know the logical value of the statement. The latter represents Kleene's third value.

T   true
F   false
I   indeterminate.

p	q	$\neg p$	$p \wedge q$	$p \vee q$	$p \equiv q$	$p \supset q$
T	T	F	T	T	T	T
T	F	F	F	T	F	F
F	T	T	F	T	F	T
F	F	T	F	F	T	T
T	I	F	I	T	I	I
F	I	T	F	I	I	T
I	T	I	I	T	I	T
I	F	I	F	I	I	I
I	I	I	I	I	$\boxed{T}$	$\boxed{T}$

The differences between the Lukasiewicz and Kleene truth tables are placed in squares in the table above.

More than Three Values

Lukasiewicz suggested a logical system that contains any number of values. For notational simplicity, truth values are represented as a set of numbers from 0 to 1. The truth value of an expression p is denoted as $\mathcal{T}(p)$. Logical operators may be defined with (Kneale & Kneale, 1984)

$$
\begin{aligned}
\mathcal{T}(\neg p) &= 1 - \mathcal{T}(p) \\
\mathcal{T}(p \supset q) &= 1 \text{ when } \mathcal{T}(p) \le \mathcal{T}(q) \\
\mathcal{T}(p \supset q) &= 1 - \mathcal{T}(p) + \mathcal{T}(q) \text{ when } \mathcal{T}(p) > \mathcal{T}(q).
\end{aligned}
$$

This produces the conventional logic when the truth values for *true* and *false* are 1 and 0, respectively. If *true* and *false* have these values and indeterminacy has the value $\frac{1}{2}$, the operators in the three-valued logic discussed above are produced.

When using a multivalued logic with over 3 possible values, modalities may be conveniently defined using

$$
\begin{aligned}
\mathcal{T}(\Diamond p) &= 1 \text{ when } \mathcal{T}(p) \ge \frac{1}{2} \\
\mathcal{T}(\Diamond p) &= 2\mathcal{T}(p) \text{ when } \mathcal{T}(p) < \frac{1}{2}.
\end{aligned}
$$

Bochvar's Logic

This three-valued logic corresponds to Kleene's three valued system with a set of weak-valued connectives proposed by Kleene but also proposed by Bochvar. There are three values: truth, falsehood, and meaningless or paradoxical statements. Moh Shaw-kwei suggested that Lukasiewicz's logic could be improved if the "middle" value were defined as paradoxical, a meaningless statement such as "This statement is false."

Using the following values,

T   true
F   false
M   meaningless,

we obtain the following truth values for each of the indicated operators:

p	q	$\neg p$	$p \wedge q$	$p \vee q$	$p \equiv q$	$p \supset q$
T	T	F	T	T	T	T
T	F	F	F	T	F	F
F	T	T	F	T	F	T
F	F	T	F	F	T	T
T	M	F	M	$\boxed{M}$	M	M
F	M	T	$\boxed{M}$	M	M	$\boxed{M}$
M	T	M	M	$\boxed{M}$	M	$\boxed{M}$
M	F	M	$\boxed{M}$	M	M	M
M	M	M	M	M	M	M

The difference between the Bochvar and Kleene truth tables is placed in a square in the table above.

Matching and Three-Valued Logic

Objects may be matched based upon more than the two logical values discussed above. The middle value may take on a number of different interpretations, including undecided (U), representing the practical inability to assign a logical value due to uncertainty, and indeterminacy (I), indicating that a logical value cannot be assigned.

Using the three-valued logic suggested by Kleene, similarity measures like the simple match and Jacquard measures can be developed. We begin with the following table and notation

	present	undecided	absent	
present	a	n	b	$a + n + b$
undecided	w	m	e	$w + m + e$
absent	c	s	d	$c + s + d$
	$a + w + c$	$n + m + s$	$b + e + d$	$T.$

This represents an expansion of the notation proposed earlier, with additional notation added for *undecided* values. In this case, m represents the middle, and n, e, s, and w represent the four points of the compass.

The simple match, given the intermediate "undecided" value, may be computed as

$$S_{1,K} = \frac{a + d}{a + b + c + d + e + m + n + s + w}.$$

This may be interpreted as the percentage of characteristics that match exactly, where matching is interpreted as meaning that the characteristic is present in both objects or missing in both documents.

The Jacquard measure may also be expanded to be consistent with Kleene's logic but suggesting a measure such as the following:

$$S_{16,K} = \frac{a}{a + b + c + e + n + s + w}.$$

This measure omits d from the numerator and the denominator as with the traditional Jacquard measure and, in addition, omits m from the denominator.

The three-valued logic proposed by Lukasiewicz has an "indeterminate" third value. Similar to the notation used for the Kleene three-valued logic, the following notion is used

	present	indeterminate	absent	
present	a	n	b	$a + n + b$
indeterminate	w	m	e	$w + m + e$
absent	c	s	d	$c + s + d$
	$a + w + c$	$n + m + s$	$b + e + d$	$T.$

The simple match, given the intermediate "indeterminate" value, may be computed as

$$S_{1,L} = \frac{a + d + m}{a + b + c + d + e + m + n + s + w},$$

that is, the percentage of characteristics that have the same value in both objects, where indeterminacy is treated as a value. Indeterminacy is treated as a legitimate value in this system; an examination of the truth tables provided earlier shows that the if and only if operator, when applied to a pair of indeterminate values, results in a *true* expression.

$$S_{16,L} = \frac{a}{a + b + c + e + n + s + w}.$$

This omits d from the numerator and omits d, and m from the denominator. The application of this third value is primarily through indexing or models expecting the assignment of indeterminate representations to topics or characteristics about which we don't have information regarding the aboutness of the document to a topic.

One could develop a measure omitting n, e, s and w, to produce the more traditional $a/(a + b + c)$ for the Jacquard measure, S_{16}.

The major drawback to the use of a three-valued logic is that the document rankings are no longer binary, with the documents themselves taking on one of three values, representing their relationship with the query. While this may be acceptable in some systems, systems based on two-valued Boolean assumptions often retrieve one of the two categories of documents, leaving the documents in the other category unretrieved. A third value complicates this; each different three-value logical model has its own interpretation of the third value, providing different motivations for choosing to retrieve or not retrieve those documents whose value may be the third, intermediate value.

The best solution to this problem is to first determine whether a search is a low-recall search, where the user wants a few good documents, or a high recall search, where the user wishes to see virtually everything on the topic. In the case of a low recall search, the documents which have an indeterminate or unknown status are best left unretrieved, with those documents with value *true* being made available for the user. In the case of a high-recall search, these indeterminate or otherwise labeled documents need to be made available to the user, perhaps with the suggestion that the query might be reformulated in such a manner that no third-valued documents are retrieved.

EXERCISE 9.5. [*Moderate*] Which multivalued logic is most consistent with the documents and queries found in text retrieval systems?

EXERCISE 9.6. [*Research Problem*] Propose a model for $\mathcal{A}$ that is consistent with Lukasiewicz's system supporting more than 3 truth values for relevance judgements.

9.10 NON-MONOTONIC REASONING

One may assume that a valid argument will remain valid if additional statements are added to the premises; this property is referred to as *monotonicity* (Genesereth & Nilsson, 1987). If $p \supset q$, then a superset of p, p_S, also implies q; thus $p_S \supset q$. Because standard logics do not allow one to produce new knowledge about the world, additions to the traditional monotonic logics are necessary if we are to make assertions that are not deductive-logically derived from the available data.

Common sense appears to be *non-monotonic* (Brewka, 1991). If one is talking about a bird, most humans will assume that it can fly and make further inferences based on this "fact." Yet, if one were to be told that the bird is a penguin, nothing has been "retracted" from the original premises; indeed, something has been added to the original system, while a conclusion has been retracted in many peoples' minds. This is clearly non-monotonic and seems intuitively to reflect aspects of the day-to-day reasoning of people. As new information is obtained (e.g. that the bird is a penguin), the set of consistent conclusions hasn't grown (non-monotonicity).

It is seldom the case that membership in a group can be precisely defined by providing necessary and sufficient conditions (Reiter, 1990). For example, almost any single part of a chair can be removed and it will still be recognized as a chair, albeit a broken one. A book might best be defined as *normally* having an author, a title, a cover, and so forth. A book might have had the cover removed. It would still be considered a book by most people, and most people would infer that the general characteristics of books apply to this mutilated book, even though it is obviously not a prototypic book.

Inferential reasoning systems, such as a forward chaining expert system, are designed to *increase* the knowledge available about a situation or environment.

New inferences result in an increase in the number of facts or inferences made. This monotonicity, an ever increasing number of inferences, is seldom seen in realistic situations where new information often functions as a corrective or partial reversal to previously inferred results. When factual information is added to a system based on traditional first-order logics, previously valid conclusions are always preserved. For previously valid conclusions to be falsified, there must be new information which contradicts previously accepted information.

This non-monotonicity can't be dealt with directly by traditional first-order predicate logic or logic based languages such as Prolog. Non-monotonicity requires extensions or modifications to conventional logics and languages.

There are several forms of belief revision that take place in real-world reasoning systems (Gärdenfors, 1992). These may include the introduction of new, non-contradictory facts, referred to as *expansion*. This is probably the easiest form of belief revision to implement. *Contraction* of the belief database requires that certain existing beliefs be removed from the set of accepted beliefs. *Revision* of the beliefs in an existing belief system requires two operations to take place: expansion and contraction. A new fact is added to the belief system, and other facts are removed through contraction.

Closed Worlds

Reiter's *closed world* assumption states that, unless specifically stated otherwise, an event or state of nature didn't occur or doesn't exist (Przymusinski, 1990; Reiter, 1980). Ramsay provides a useful statement of the closed world assumption: "if you can't prove P, assume $\neg P$" (Ramsay, 1988). This provides a formal mechanism for the representation of logical negation which is lacking from most traditional first-order logics.

The closed world assumption is very effective in environments where relatively few of the possible predicate-argument pairs are explicitly mentioned. Let us assume that we have the following statements about book 17:

- $About_{17,dogs}$

- $About_{17,cats}$

- $About_{17,rabies}$

It might be helpful in many retrieval applications to explicitly assume that if a book isn't explicitly asserted to be about a specific topic than it isn't about the topic. We might then assume $\neg About_{17,broccoli}$ and $\neg About_{17,Pennsylvania}$

Allowing all possible predicates to be represented in a negative form if not explicitly asserted may result in contradictions and an explosion in the number of predicates. One may choose to limit the negations to a specific predicate, such as the "about" predicates.

Default Logic

In many situations, humans appear to assume certain attributes of items and predicates (Ramsay, 1988; Przymusinski, 1990). Logical values may be assumed unless known to be false because of other evidence. Defaults allow attributes or predicates associated with other entities to be implied by using the entity. Formally, accepting $\alpha(x)$ as true and $\beta(x)$ as a default, one can conclude $\beta(x)$ if $\neg\beta(x)$ is not believed or accepted. For example, a *chair* might have, as one of its defaults, *legs*. Conclusions can be drawn about the *chair* assuming the presence of *chair legs* unless information is available explicitly precluding the presence of *chair legs*.

When using default logic, traditional rules of formal logic may be applied. Given a proposition such as $p \supset q$ where q is a default, we assume q to be true unless there is evidence otherwise. Thus, if p is true, the rules of formal logic allow us to deduce q. If p represents "x is a bird" and q represents "x can be consistently assumed to fly", and if one accepts p, q logically follows *unless* it is inconsistent with the available facts (such as that penguins don't fly).

A default logic based theory may be understood as a set of true statements, the theory, and default rules, which allow for information to be generated that could not be generated by the "theory" alone. The set of statements consistent with both the true statements and the defaults is referred to as the *extension*.

Autoepistemic Logic

Autoepistemic logic concerns itself with formalizing a deductive logic providing conclusions when there is no evidence conflicting with defaults (Moore, 1985; Przymusinski, 1990). This function is obtained by proposing a belief function $\mathcal{B}$, where inclusion of x in the set of beliefs is denoted as $\mathcal{B}(x)$. There is a strong similarity between $\mathcal{B}(x)$ and the modal logic statement $\neg\Diamond\neg x$, it is not possible that x is false. Other definitions of belief are possible, but are less adequate than than $\mathcal{B}(x) = \neg\Diamond\neg x$. Other possible definitions of belief are $\mathcal{B}'(x) = x$ and $\mathcal{B}''(x) = \Diamond x$. The former suggests that the truth value of belief in a fact is the same as the truth value of the fact itself. The latter suggests that the truth value of believing in a fact is the same as the truth value of the possibility of the fact.

The assumption that a *chair* has *legs* unless believed otherwise may be represented by

$$\forall x\,((chair(x) \wedge \neg\mathcal{B}(\neg legs(x))) \supset legs(x))\,.$$

Thus, unless one believes otherwise, it is assumed that *chairs* have *legs*.

The relationships between forms of epistemic logic and default logic have been examined (Reiter, 1990). The extensions of specific models of the two can be shown to be identical, suggesting that the strong intuitive appearance of similarities between the two can be formally justified.

Circumspection

Problems are often implicitly circumscribed (McCarthy, 1980; Przymusinski, 1990; Reiter, 1990). The use of factors not mentioned in the problem are often implicitly rejected. In the classic missionaries and cannibals problem, a boat is always posed as available for transporting up to 3 passengers to either side of the river. The problem solver is expected to not "cheat" by posing a bridge down stream, the availability of helicopters, and so forth (Ramsay, 1988). Problems are usually stated in a positive form, suggesting what may be used in solving the problem, as opposed to what may not be used.

The circumscription model suggests that there might be two levels of goal satisfaction. The properties of a difficult to achieve and precisely defined goal, G, may be approached by explicitly attempting to satisfy a more general goal G' with properties P'. A solution that satisfies G' will satisfy G if G' entails G.

Document filtering using logic requires extensive use of such non-monotonic logics both because of the ever increasing knowledge in the form of modifications of existing knowledge, and because of constant feedback provided by the user(s) (Hurt, 1998). Much more research is need in the application of non-monotonic logics to information retrieval problems and the analytic modeling of these systems.

9.11 SUMMARY

Formal logics are useful as filtering and retrieval languages, providing lists of those characteristics to be found in documents passed by the filters. As a language capable of expressing causal relationships, logic is a powerful tool that has been widely studied and has been suggested to be an appropriate tool for modeling many real-world scenarios. Analytic tools can be applied to some filtering systems consistent with logical considerations to yield retrieval performance statements. Using these analytic methods, potential strengths and weaknesses of different logical methods can be understood in light of their effect on retrieval performance.

10 LINGUISTIC KNOWLEDGE

He multiplies words without knowledge.

—Job 35:16

10.1 INTRODUCTION

Natural language processing contributes to improved retrieval performance by extracting from natural language text information about terms and their relationships. This information is far richer than what is obtained with term frequency methods that assume the statistical independence of terms. Once acquired, this linguistic knowledge may be reflected in retrieval and filtering systems in either a modified query or in a modified document incorporating this information. *We can expect natural language processing to improve filtering and retrieval performance if, and only if, the application of linguistically derived information increases the ability of the retrieval system to discriminate between documents of differing relevance.* While linguistic knowledge may be obtained through purely statistical analysis, humans may extract this same information without using massive number crunching capabilities, and it is likely that, for automated systems, linguistic methods may be ultimately simpler and faster at extracting information that improves retrieval performance than are methods that explicitly incorporate higher order statistical dependencies. We present a model of grammatical parsing and part-of-speech tagging that allows us to make specific claims about the level of retrieval and filtering performance that will be obtained when linguistic knowledge is incorporated. The model provides both upper and lower bounds for performance with the best-case and worst-case part-of-speech tagging.

The purpose of this examination of how natural language processing improves retrieval and filtering performance is to

- determine expected performance,

- make precise statements about what complex of linguistic structures and applications result in superior filtering performance, and

- learn how individual linguistic factors affect performance.

We can accomplish these goals through the application of the analytic models discussed earlier. This chapter is an extension of Chapter 8, which addressed the performance of systems incorporating statistical relationships between terms.

Most research examining either linguistic phenomena or information retrieval and filterings systems, separately or in conjunction, describe system performance by examining average retrospective performance figures, or the characteristics of specific instances or phenomena. However, exact *expected* results may be obtained through modeling enough aspects of a filtering system to provide information about the expected characteristics of the system. This is different than computing a single average for an experimental data set. The analytic approach allows us to formally understand how linguistic phenomena interact to produce a certain level of performance. Analytic techniques complement experimental methods and empirical data-gathering methodologies. While the impact of some assumptions, as well as some models of natural language processing, may best be studied empirically, many other questions addressing the relationship between natural language processing and information filtering are best answered analytically.

The purpose of applying linguistic knowledge to documents and queries in filtering systems is to provide additional information about individual terms and sets of terms beyond the simple frequency of term occurrences. This occurs primarily through two mechanisms: part-of-speech tags (e.g. *noun, verb,* or *subject*), and structure producing phenomena, such as grammatical parsing, that can be used to identify relationships between terms. For example, the phrase representing the subject of the sentence, *The brown spot was on the green shirt,* refers to a brown spot. Identifying the spot as brown, as opposed to green, requires a parsing procedure which shows that the relationship between brown and spot is different than the relationship between spot and green and that *the green shirt's brown spot* has a different and what intuitively seems to be a stronger relationship between *green* and *shirt* than between *brown* and *shirt.*

To design and understand retrieval and filtering, we need to understand how the output of linguistic processes affect text filtering and retrieval performance through statistical dependencies between terms, as well as through the choice of terms and attached part-of-speech tags. Terms in natural language may carry information that may be elicited more easily and more accurately through linguistic rather than statistical techniques. Whether one approach or another is better in some cases, or in all cases, can be determined analytically, given accurate parameter values.

The basic question examined in this chapter is

under what circumstances will incorporating linguistic knowledge result in superior filtering performance, and under what circumstances will performance grow worse when linguistic techniques are applied?

By providing analytically based statements of when a linguistic method improves retrieval performance, combined with the results of earlier chapters, we can answer this question. If the ability to filter information in a retrieval or network application is reflective of what allows humans to filter information and retrieve facts, our results may be indicative of the utility of linguistic methods for humans in all domains, not just filtering and retrieval.

10.2 TAGGING AND SUFFIX STRIPPING WITH ONE TERM

Grammatical part-of-speech tags are assigned in a computational environment though the *tagging* process (Brill, 1994). Tagging may be performed by a tagging program that operates based on either grammatical rules that are programmed into the system or learned, or tags may be assigned based upon statistical considerations, often using Hidden Markov Models. Assigning a part-of-speech tag to a term token allows users to discriminate between different senses in which a term is used. A part-of-speech tag assigned to the term *girl* in sentences like *girl bites dog* and *dog bites girl* can be used by a retrieval system to retrieve only those documents in which a girl acts as biter (subject) or those documents in which the girl is bitten (object). While using natural language processing techniques such as tagging may improve retrieval and filtering performance, the degree of improvement with word-sense disambiguation may vary from small to moderate amounts (Burgin & Dillon, 1992; Krovetz & Croft, 1992; Sanderson, 1994; Strzalowski, 1995). The analytic rules described below can lead to an understanding of how tagging contributes toward the matching of queries and text.

We initially make some simplifying assumptions to allow us to understand the nature and benefits of part-of-speech tagging. For example, we limit ourselves to queries with a single term to allow us to examine the impact of tagging on a single term, taken in isolation. We thus treat a single (query) term and its tag together as a single complex. We also assume optimal retrieval so that we can avoid the complexity added by working with a sub-optimal ranking method. Earlier work on analytic prediction of retrieval performance shows how a model may assume multiple terms (Chapter 8) and sub-optimal retrieval (Chapter 5), but such complex models may conceal some simple underlying phenomena.

The basic unit of linguistic analysis is the *term* or word. We refer to an occurrence of a term as a term *token* and the generic form for the term as the

Table 10.1.   Sample documents, where "Y" denotes *yes* and "N" denotes *no*.

Relevant Documents		Non-relevant Documents	
Term Present?	*Query Tag?*	*Term Present?*	*Query Tag?*
Y	Y	Y	N
Y	Y	Y	Y
Y	N	N	N
N	N	N	N
N	N	N	N

term *type*. The distinction is roughly that between numeral and number (the former corresponding to token, the latter to type).

What constitutes a term is open to debate; we follow the convention that text strings that are bounded by spaces are terms. This crude definition is adequate in most circumstances, although questions are not addressed such as whether more complex expressions such as *take-it-or-leave-it*, *Greco-Roman*, or *computer-human interface* are best treated as one term or two, or whether each phrase is a single or multiple concept. Phrases contain groups of terms that together address a particular concept or construct. Sentences are then composed of one or more phrases representing an entire statement. Parsing decomposes a sentence into phrases and individual terms occurrences, each with part-of-speech tags attached. The output of a parser, a *parse tree,* is a hierarchy of tags assigned to a sentence, with tags being assigned to individual terms as well as to phrases. Phrases and entire sentences are discussed in later sections of this chapter.

Part-of-Speech Tags and Single Term Performance

Understanding how part-of-speech tags can affect retrieval performance can lead one to make better decisions about the use of natural language processing in filtering and retrieval systems. When tagging query terms in documents, the number of documents with the tagged query term will be the same as or fewer than would be the number of documents with the untagged term if no tagging were used. As before, the probability that the query term occurs in a document is denoted as t, and the probability that a document has the query term, given that the document is relevant, is denoted as p. The probability that a document is tagged with the query tag, given that it has the query term, is denoted as τ. The probability that a document has the query term and is tagged with the query tag is the product $t\tau$. Similarly, the probability that a term is tagged with the query tag, given that the document has the term and is relevant, is π. The probability that a relevant document contains the term tagged with the query term's tag is the product $p\pi$.

Theorem 10.1 (Tagging improving performance) *Given optimal ranking, performance is improved if and only if*

$$1 + t\tau - p\pi < 1 + t - p. \tag{10.1}$$

Here the left hand side represents $\mathcal{A}$ with tagging and the right hand side $\mathcal{A}$ without tagging, where $\mathcal{A}$ (Equation 6.1) is the expected position of a relevant document in the ordered list of documents, scaled to be in a range from 0 to 1. A term is assigned a part-of-speech tag with the expectation that the tagging will increase retrieval or filtering performance. Filtering performance is improved with tagging if and only if the ASL with the tagging is less than the ASL without the tagging. This condition is met when the $\mathcal{A}$ with tagging is less than the $\mathcal{A}$ value without tagging.

Corollary 10.1 (Tagging decreasing performance) *Tagging decreases retrieval performance when*

$$1 + t\tau - p\pi > 1 + t - p.$$

We can measure the degree to which retrieval performance with tagging exceeds the performance without tagging by reformulating Equation 10.1 so that a Tagging Improvement Factor (TIF) represents the amount added to the left hand side of Equation 10.1 by tagging, with a positive value for the TIF indicating that tagging improves performance (decreases ASL), and a negative value indicates that tagging decreases system performance.

Theorem 10.2 (Tagging improvement factor) *The* Tagging Improvement Factor *(TIF), the improvement in $\mathcal{A}$ when part-of-speech tagging is used, is computed as* $TIF = t(1 - \tau) - p(1 - \pi)$.

This follows from

$$1 + t\tau - p\pi + \text{TIF} = 1 + t - p \tag{10.2}$$

and then

$$\text{TIF} = t(1 - \tau) - p(1 - \pi). \tag{10.3}$$

The TIF is thus computed as the proportion of all documents with the term in question that aren't tagged minus the proportion of relevant documents with the term in question that aren't tagged. A TIF is 0 when the same proportion of all documents as relevant documents are assigned the tag in question. This occurs when the tags are distributed the same in both the set of relevant documents and in all documents. If a tag occurs with greater relative frequency in the relevant documents, the tagging results in improved performance.

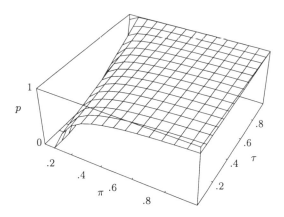

Figure 10.1.   The break-even π points for deciding to tag or not tag a term, with $t = 1/10$.

Corollary 10.2 (Change in ASL due to tagging) *The change in ASL for a single term due to part-of-speech tagging may be computed as*

$$ASL_\Delta = N\left(\mathcal{Q}(A - TIF) + \overline{\mathcal{Q}}(\overline{A} + TIF)\right) \tag{10.4}$$

where ASL is the expected position of relevant document (Equation 6.5) and $\mathcal{Q}$ is the probability the ranking used is optimal (Equation 5.1).

Example.   The data in Figures 10.1 and 10.2 show the break-even points between text filtering performance with and without part-of-speech tags, that is, where the TIF is 0. Giving varying values of p, τ, and π, and $t = 1/10$ for Figure 10.1 and $t = 1/2$ for Figure 10.2, we find that, for lower values of t, the p value rises to both higher values and rises more quickly than for higher values of t. The area representing positive TIF values (improved performance with tagging) is above the surface, with area below the surface representing values that will decrease performance. The range of values that will result in improvements with tagging increases with t.

When tagging is applied to the data in Table 10.1, where $\tau = 1/2$ and $\pi = 2/3$, we compute the ASL as $\frac{10}{2}(1 + \frac{1}{2}\frac{1}{2} - \frac{2}{3}\frac{3}{5}) + \frac{1}{2} = 4.75$. If one examines Figure 10.2 one finds that for $t = 1/2$, $p = 3/5$, $\pi = 2/3$ and $\tau = 1/2$, we are at a point above the break-even surface but not very far from the surface, indicating that tagging helps somewhat in this case.

Figure 10.3 shows the ASL when $t = 1/2$ and $p = 3/5$ for both a tagged query (the fine mesh) and for an untagged query (with larger holes in the mesh). Figure 10.4 similarly shows the ASL when $t = 1/10$, with everything else similar except for the ASL. For the case where there is a low t and a large

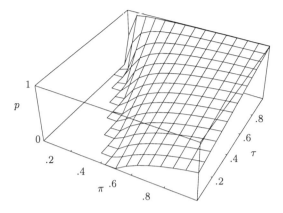

Figure 10.2.   The break-even π points for deciding to tag or not tag a term, with $t = 1/2$.

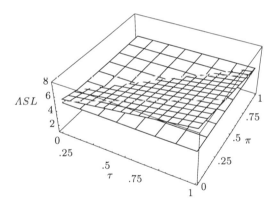

Figure 10.3.   The Average Search Lengths (ASL) for two searches of ten sentences. The large mesh (large holes) represents retrieval with no tags and with $t = 1/2$ and $p = 3/5$. The smaller mesh represents a search with tagging, where the parameters are as above but where π and τ are allowed to vary.

gap between t and p, the terms are already good discriminators and the region where tagging will improve performance (the top part of the figure where the large mesh is above the finer mesh, the latter representing tagged performance) is small. For this data, tagging results in improved performance with very high π and, to a lesser extent, for lower τ.

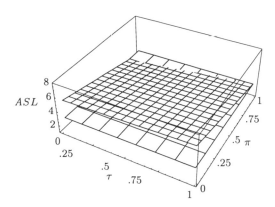

Figure 10.4.   The Average Search Lengths (ASL) for two searches of ten documents. The large mesh (large holes) represents retrieval with no tags and with $t = 1/10$ and $p = 3/5$. The smaller mesh represents a search with tagging, where the parameters are as above but where π and τ are allowed to vary.

Best and Worst-case Performance with Tagging

Retrieval performance may be viewed as being dominated by t and p. Part-of-speech tagging may be viewed as providing a means of improving on this by modifying these values through τ and π. However, part-of-speech tagging cannot overcome some limits imposed by the untagged probabilities. Note that our initial development assumes very large databases and associated approximations; later, some exact values will be computed which will be accurate with small databases.

Theorem 10.3 (Approximate best-case tagging performance) *The best performance obtained with tags is*

$$ASL = \frac{N}{2}(1 - p) + \frac{1}{2} \tag{10.5}$$

and the corresponding best-case $\mathcal{A}$ is $1 - p$.

The highest TIF (Equation 10.3) occurs when the proportion of all documents with the query term tagged as in the query approaches its minimum ($\tau \to 0$) and the proportion of relevant documents with the query term tagged as in the query approaches its maximum ($\pi \to 1$). This approaches,

$$ASL = \frac{N}{2}(1 + 0t - 1p) + 1/2 = \frac{N}{2}(1 - p) + \frac{1}{2}.$$

Corollary 10.3 (Approximate worst-case tagging) *The worst-case performance obtained with tags is*

$$ASL = \frac{N}{2}(1 + t) + \frac{1}{2} \tag{10.6}$$

and the corresponding worst-case $\mathcal{A}$ is $1 + t$.

The worst case occurs when $\tau \to 1$ and $\pi \to 0$. In this case,

$$ASL = \frac{N}{2}(1 + 1t - 0p) + 1/2 = \frac{N}{2}(1 + t) + \frac{1}{2}.$$

We can see that performance will be proportionally no better than would be obtained with an $\mathcal{A}$ factor of $(1 - p)/2$ and no worse than $(1 + t)/2$. The range of retrieval performance is thus bracketed between that obtainable with optimal tagging and that with the worst case tagging.

In reality, τ doesn't approach very close to 0 with small databases, such as in Table 10.1, although τ does approach very close to 0 with larger databases. With smaller databases, or when a close approximation isn't satisfactory for a large database and it is desirable to compute the exact value, it becomes necessary to take into account how close τ is to 0 when computing the best-case and worst-case performance values. This is because the limits described earlier are not met (by a factor gp, where generality $g = \Pr(rel)$).

Theorem 10.4 (Exact best-case tagging performance) *The exact upper performance bounds with part-of-speech tagging and generality g are*

$$ASL = \frac{N}{2}(1 + gp - p) + \frac{1}{2}. \tag{10.7}$$

This assumes that $\pi = 1$ and $\tau \to 1$ as in Equation 10.5. There must be at least (gp) tagged terms in the document database. The gp values in Equation 10.7 becomes very small, approaching 0 when a very large, realistic database is used, so that this equation in the limiting case approaches Equation 10.5.

In most realistic searches, t will be rather small if it is a good search term, usually below .01. When p is much higher than this, as would be the case with a strongly discriminating term, the potential for improvement with tagging is far better than the potential for a decrease in performance.

Using the data in Table 10.1, $N = 10$, $p = 3/5$, $t = 1/2$, and $g = 1/2$, and using the approximate best and worst case measures described in Equations 10.5, and 10.6, $8 \geq ASL \geq 2.5$. Using the exact formula (Equation 10.7) however, we find the $ASL \geq 4$. The ASL computed earlier for this tagged query was 4.75. It is clear that this is better than the ASL of 5 found when no tagging occurs but is not as good as it could be, not equaling the best-case performance (4). Note that random retrieval here would result in $ASL = 5.5$.

Construction of Optimal Tags

Continuing with this example, optimal or best-case tagging and the ASL of 4 could be obtained here by tagging all relevant documents with the term in Table 10.1 and by not tagging those non-relevant documents with the term. In this situation, the three relevant documents with the term would be at the beginning of the ranked list (average position 2) and the other two relevant documents would be at the average position of 7, producing $ASL = 4$.

This suggests the following:

Theorem 10.5 (Constructing optimal tags) *Given that the upper bounds of performance are obtained as with Equation 10.7, where the active component of A is $1 + \tau t - \pi p$ and where $\tau \to 0$ and $\pi \to 1$, the best-possible tagging may be constructed by tagging all relevant documents containing the term with the query tag and none of the non-relevant documents with the term with the query tag.*

Clearly, this has the effect of maximizing π and minimizing τ, subject to the constraint imposed by g which is assumed to be fixed above. We see that there must be at least gp documents with the tagged term; if these are all relevant, then the upperbounds are obtained.

Stemming and Suffix Removal

Stemming terms, that is, removing suffixes from terms, may be seen as a form of "untagging" of terms, finding a commonality between this term and another term by removing information that previously made a term unique. Stemming may improve performance (Harman, 1991; Popovic & Willett, 1992; Porter, 1980) by increasing recall (Kraaij & Pohlmann, 1996). Stemming terms such as *stemming* can produce a stem such as *stemm* or *stem*. In a sense, it is not important whether the stem is that which would be produced by a human; what is important is that the stems be consistent, so that if one were looking for works on *stemming*, one would find them.

Terms with stems correlate weakly with the stemmed terms, suggesting that there is a relationship between the root and the unstemmed version (Church, 1995). This correlation may approach one half for terms that are good discriminators and approaches zero for terms that are poor discriminators.

Different stemming methods yield different levels of performance, with the relative performance levels depending significantly on the choice of performance measure used; which is the best stemmer procedure for a given situation is still an open question (Paice, 1994).

Using the inverse of the methods described above, the performance of systems using stemming in a single term case may be studied. Modifying Equation 10.2, we may say that the A factor without stemming (with suffixes and thus similar to tagging with part-of-speech tags) is $1 + \tau t - \pi p + c$, while, with stemming (untagging is similar to suffix removal), the A factor is $1 + t - p$, where c is the Suffix Decreasing Factor (SDF), the decrease in performance due to c set as in Equation 10.3. The best and worst case performance figures without stemming (with tagging) are as in earlier discussions about part-of-speech tagging; their development is posed as an exercise below.

EXERCISE 10.1. [*Easy*] Assume that there are 100 documents, half with the term, and 50 of the documents are relevant, 40 of them with the term. When considering the tagging of terms, the term (with the query tag) is in 40 of the documents and in 35 of the relevant documents. What is A with and without tagging? What is the ASL with and without tagging? Does part-of-speech tagging help or hurt performance in this case?

EXERCISE 10.2. [*Moderate*] What are the exact lower bounds for retrieval performance with tags (the complement to Equation 10.7)?

EXERCISE 10.3. [*Difficult*] Develop a full set of theorems and conjectures for stemming that are similar to those developed for part-of-speech tagging.

10.3   MULTIPLE TAGS PER TERM TYPE

Filtering performance can obviously be enhanced through the use of supplemental part-of-speech tags. It is also clear that tagging can decrease performance. While the preceding section addressed the use of single tags with single term queries, it is helpful to consider performance when a single term in an untagged environment can explicitly produce two different tagged terms, one tagged as with the query and the other tagged differently. For example, the untagged term *run* may be labeled as either a verb, such as would occur in statements such as *Fred will run for president* or *Caitlyn will run after the cat,* while *run* labeled as a noun would be found in statements such as *The dog was on its run, Jim participated in the 10K fun run,* and *June got a run in her stocking.* Consider the situation where a single term such as *run* can become more than one possible tagged term when tagged, and each of the terms is to be used positively or negatively by the retrieval or filtering system.

As an example, consider a textile manufacturer who is searching for studies on how to stop runs in synthetic fabrics. One system has untagged terms, and *run* occurs with probability .05 in all documents and .2 in relevant documents. We can thus compute $A = (1-.2+.05)/2 = .425$. A second database, containing exactly the same documents and has terms tagged as to whether they are nouns or verbs. The term complex *run-noun* occurs with probability .05 in all documents and probability .5 in relevant documents, thus $A = .37$. The term complex *run-verb* occurs with probability .05 in the set of all documents and probability of .01 in relevant documents, thus $A = .57$.

In this case, one A for a tagged term is better (lower) than the A for the untagged term, while the other A for a tagged term is worse than the A is for the untagged term. An additional method useful for studying A, when there are a number of queries, is to examine the expected A:

$$E(A) = \sum \Pr(i)A_i$$

where i is taken here over the set of possible parts-of-speech. We limited ourselves to one untagged term. In the preceding example, if *run-verb* occurs with probability .7 and *run-noun* occurs with probability .3, then $E(\mathcal{A}) = .3(.37) + .7(.57) = .51$

Definition 10.1 (Effective tagging) *If the expected $\mathcal{A}$ for a set of taggings is less that the $\mathcal{A}$ for the untagged terms, then the tagging is said to be* effective.

If $E(\mathcal{A})_{tagged} < \mathcal{A}_{untagged}$ (in our example, $.51 < .425$) then tagging is superior, on the average, to having untagged documents and queries. If the inequality doesn't hold (as it doesn't hold in our case), then tagging will provide inferior results, on the average.

Entries like part-of-speech tags may be applied to other types of tagging or additions to terms and documents. For example, the addition of metadata tags can allow type-of-document information to be used in retrieval. One can also use controlled vocabulary terms to improve retrieval performance.

EXERCISE 10.4. [*Research Problem*] Consider several term types, each of which may have several different part-of-speech tags. How does performance vary with the presence or absence of these different tags?

EXERCISE 10.5. [*Research Problem*] Using the results of the previous exercise, how do phrase tags as well as individual part-of-speech tags affect performance. Consider for example a tag *noun phrase*, for a sequence such as *determiner* followed by *noun* for the phrase *the book*. Thus, what is the performance with individual term tags such as *determiner* and *noun* versus performance with these tags and the phrase tags such as *noun phrase*?

10.4 CONTROLLED VOCABULARIES

Documents and multimedia may be represented using any of the naturally occurring features in each document, or the representation may be features selected from a set of *controlled* terms. A controlled vocabulary is a set of terms or phrases that contains unique ways of representing concepts occurring in the database, avoiding the "crisscross of many-one and one-many relationships between words and their referents" (Svenonius, 1986). Some have viewed the user of controlled vocabularies as enhancing the precision and recall of a search (Svenonius, 1986) while others have suggested that in some environments a controlled vocabulary must be viewed as a "precision device and not that of a recall device" (Boyce & McLain, 1989).

Which vocabulary type performs best has long been the subject of experimental work, but it can be addressed analytically as well as experimentally. Deciding when to use one vocabulary type or another may thus be studied by noting parameter values for different databases and the nature of searches, noting whether or not the conditions of the analytically determined rules are met.

For our analytic work below, we will usually address individual terms or phrases rather than groups of terms or multiple phrases. Thus, when studying a query such as *dogs and rabies* the relationships and dependencies between these two concepts will be ignored. Put more positively, the terms will be treated as though they are statistically independent. The separate concepts will be treated separately, and the performance will be studied for queries as though there was a single term or phrase in each query, thus allowing us to isolate problems. One can quickly lose sight of underlying mechanisms, and our focus here is explicitly on understanding when controlled vocabularies provide better representations for retrieval than are obtained with uncontrolled vocabularies.

Controlled terms can be used to represent equivalent or similar meanings, different spellings or abbreviations or translations, as well as hierarchical relationships. These concepts have multiple ways they can be represented by a single text string in a controlled vocabulary. The binary term frequency of a controlled vocabulary term d_c in a document may be computed as the maximum of a set of n other uncontrolled synonyms in a document, thus $d_c = max(d_1, d_2, d_3, \ldots, d_n)$. This assumes that any occurrence of d_i, $i = 1, 2, \ldots, n$ represents an occurrence of the idea underlying d_c. The controlled vocabulary expression *house pets* might be assigned the maximum binary frequency in a document from the uncontrolled terms *dogs, cats, tropical fish*, and *parakeets*.

Homonyms (and more properly homographs) are words with different meanings that sound (or are written) the same. The distinction between the two is important, as *two, too,* and *to* are homonyms and sound the same but are written differently. Homographs may best be understood as a written text string with more than one definition or meaning. When a searcher wishes to find a particular meaning or concept, it may be necessary to retrieve all documents containing the meaning, as represented by a particular term, and all the documents containing homographs for that term. Searching using a controlled vocabulary would help the user exclude non-relevant documents containing homographs.

Performance with Controlled Vocabularies

We can easily describe the performance of retrieval with a single term or concept. Given a choice of using one controlled vocabulary term (or a phrase, which may be treated as a single unit) or a single free-text, uncontrolled term, which should be used? Humans use their own rules for searching, with professionals using more sophisticated methods that take into account many variables and nuances not addressed by an analytic model, resulting in superior retrieval performance (Fidel, 1992). The following rule can be used by searchers to improve performance:

> Choose to use a controlled vocabulary term if and only if the performance using the controlled term is better than that obtained with the free-text, uncontrolled term.

Performance with a controlled vocabulary query is superior to performance with an uncontrolled vocabulary query when

$$\mathcal{A}_c \leq \mathcal{A}_u, \qquad (10.8)$$

where A_c denotes the A value for controlled vocabulary and A_u denotes the A value for uncontrolled vocabulary.

Theorem 10.6 (Controlled vocabulary & performance) *Controlled vocabulary improves performance when*

$$t_c - p_c \leq t_u - p_u,$$

where the subscripts u and c represent uncontrolled and controlled vocabulary parameters, respectively.

A transformation may illuminate the problem further:

Corollary 10.4 *A controlled vocabulary will outperform an uncontrolled vocabulary when*

$$t_c - t_u \leq p_c - p_u$$

that is, performance will be better using a controlled vocabulary when the difference between the rate of occurrence of the term between documents with controlled vocabularies and terms without controlled vocabularies is less than the difference between the rates of occurrences (in relevant documents only) for the two vocabulary types.

When considering the entire controlled vs. uncontrolled vocabulary, instead of just a single term, one may wish to use a controlled vocabulary only when, on an average, it decreases $\mathcal{A}$ when compared to what is obtained with uncontrolled vocabularies. This may be expressed as

$$E(\mathcal{A}_c) \leq E(\mathcal{A}_u). \tag{10.9}$$

EXERCISE 10.6. [*Moderate*] Arbitrarily select one term related to hobbies or forms of entertainment you enjoy. Find the controlled vocabulary counterpart for this term by looking up the term in the list used to provide controlled vocabulary for a library or indexing service, such as the *Library of Congress Subject Headings*. If the controlled and uncontrolled terms are the same, start over. Search an online catalog for all occurrences of documents with either the controlled or the uncontrolled terms and judge their relevance. Using Theorem 10.6 and parameter values for this dataset, will the controlled vocabulary perform better than the uncontrolled vocabulary?

EXERCISE 10.7. [*Research Problem*] Assume that there is a single underlying meaning for all occurrences of any term within a set of terms that are synonymous with each other. Treat this underlying meaning as a tag of sorts. Provide conditions under which the use of synonymous terms increases or decreases performance. Under what condition would using synonymous terms results in performance inferior or superior to that provided by a single controlled vocabulary term?

10.5 MULTIPLE TERMS AND GRAMMATICAL STRUCTURES

Instead of using statistically independent individual terms when computing document probabilities, term pairs, triples, etc. may be used. The occurrence of n consecutive linguistic units, such as terms, syllables, or letters, may be treated as a unit referred to as an n-gram. Computing the probabilities for these units may be valuable in that structural information is statistically captured. Implicit in limiting the system's focus to windows of a particular size is that one term is more likely to be related to a neighboring term than it is to be related to a distant term (Haas & Losee, 1994; Jacquemin, 1996).

Neighboring terms may be treated so that the order of the terms in the data is considered or so the order that naturally occurs is ignored when computing probabilities. The latter is usually the case in statistical information retrieval models. Treating terms without regard to their order will allow for more accurate estimates to be made, but this is particularly difficult with larger n-grams (Losee, 1994a). Several methods are available for incorporating dependencies or term ordering, including the Bahadur Lazarsfeld Expansion, Bayesian networks, Hidden Markov Models, and autocorrelations, as discussed in Chapter 7.

The probabilistic relationships between terms or phrases may be estimated from the parts-of-speech of the terms in question, as well as from the parts-of-speech of neighboring terms. Different pairs of grammatical components will have different correlations and knowledge of these may be used as estimators of the degree of association between two specific words. If, for example, there is a probability of .25 that an adjective is followed immediately by a noun, this might be used as an estimate of the relationships between *ecru* and *spaceship*, a term pair that may occur with such infrequency that the accurate estimation of the relationship between the two terms is almost impossible.

Non-topical characteristics such as writing *style* may be learned statistically, given the availability of data containing this information. Learning the statistical relationships between grammatical components, as represented by part-of-speech tags, can allow a system to capture knowledge about structural relationships. Consider prose such as:

See Jane run.
Run Jane, run.
Watch Spot go.
Go Spot, go.

This is easily identified by most adults as rather boring prose written for beginning readers. The grammatical structure here is highly regular, and the reader could probably develop a simple parody of this kind of prose for almost any subject matter in a few moments. These grammatical patterns are easily represented statistically, and when one wishes to retrieve documents of this grammatical type, one can discriminate using probabilities describing the structures in relevant documents and in all documents. Similarly, one can discriminate based on a wide variety of non-topical aspects of documents such as linguistic characteristics and semantic styles. The more subtle are the aspects, the more data is required from which to learn with a given degree of discriminatory accuracy.

Some grammatical constructs occur with a higher rate when using discipline-specific vocabularies than when using more general vocabularies (Losee, 1996). If we believe that these constructs represent those portions of a document that are most likely to be found relevant by the user, then structures with these correlations may provide estimators for correlations between subject-bearing terms in relevant documents. The expected mutual information measure (EMIM) may prove computationally useful when describing these inter-term relationships. Remember that it is otherwise difficult to accurately estimate probabilities in relevant documents for complex structures.

EXERCISE 10.8. [*Difficult*] The adjective *ecru* occurs in 2 of 4 relevant documents that have been retrieved and the noun *spaceship* occurs in 3 of the 4. If the correlation between an adjective and a following noun is 1/4, what is a reasonable estimate of the probability of the phrase *ecru spaceship* occurring in a relevant document?

Frames, Slots, and Cases

To study groups of terms in useful relationships, viewing the term clusters as *frames* (Metzing, 1979) or *cases* (Fillmore, 1977) or *case frames* (Charniak & McDermott, 1985) may prove helpful. These structures provide a means of representing the structure or meaning inherent in natural language in a standardized and structured way. A case or a frame represents a scenario, a set of events, actions, or places that can be described. A frame contains *slots*, representing characteristics of the scenario. Slots may have values derived from a natural language statement or may have default values. For example, a frame for the Christmas scenario might have a default established that if there is a *Christmas tree*, then this tree is assumed to be *coniferous*, *decorated*, and *under twenty-feet tall*.

Assume that a query may be treated as a single frame or case. Retrieval may be based upon either an exact match between the query and a case representing a document, with the matching being between entire frames or specified parts of frames. The document cases may be ranked based upon probabilistic considerations, with the query case providing initial knowledge about the parameters of query features.

Frames also may be ranked for presentation to the searcher in order of their retrieval status value, associated with the degree of relationship between the query and the document frames. Different slots within the frame may have different discrimination values, depending upon the relationships between slot value frequencies in the slots and the query.

The probability of a particular frame with a given set of slot values may be computed from historical data. This probability may be decomposed into two probabilities, the probability that the type of frame found is the type that did occur, and the probability the frame will have the values that it has, given that it is this type of frame that has been found. The product of these two probabilities is the unconditional probability that a particular frame occurs. The probabilities associated with frames and cases may be used to compute the expected performance. More traditional grammars may also be used in computing expected performance, as is discussed below.

10.6   THE STRUCTURE OF STATEMENTS

Much of modern linguistic theory is based on grammars taken from the Chomsky hierarchy of types of formal languages. Grammars represent the structure of sentences and the relationships between terms and are thus important components in the production of relationships that affect retrieval performance. Chomsky, who argued in 1969 that "It must be recognized that the notion of a 'probability of a sentence' is an entirely useless one, under any interpretation of this term" (Ney, 1997), moved the assignment of part-of-speech tags away from the statistical, although tagging has been moving back toward the statistical as a result of recent successes. For Chomsky, a grammar is a 4-tuple $G = < T, z, \mathcal{R}, S >$ where T is the set of non-terminal symbols in the language, e.g. $A, B, \dots , Y$, the set z is the set of terminal symbols, e.g., $a, b, \dots , y$, m grammatical rewrite rules $\mathcal{R} = r_1, r_2, \dots , r_m$, and S is the starting symbol.

A rule might be written as $A \to BCD$, indicating that non-terminal symbol A is replaced by ("rewritten as") the non-terminal symbols B, C, and D, in that order. Non-terminal symbols such as A and B are often rewritten eventually as terminal symbols, tokens occurring in a sentence.

The Chomsky hierarchy is a ranking of grammar types, with each type being precisely described by a set of constraints on grammars of that type. A grammar of type n produces or describes a language of type n. A type n language is also a language of type $n - 1$. As one moves from a type n grammar to a type $n - 1$ grammar, the restrictions on the grammar are relaxed, with a type 0 grammar having the fewest restrictions.

Regular (Type 3) Grammars

The most restrictive set of rules are assigned to a *regular finite state*, or *type 3* grammar. The right hand side (RHS) of each rule contains one terminal symbol and optionally a non-terminal symbol. Thus, a type 3 grammar allows the replacement string in a production to be restricted to either terminal symbols or a terminal symbol immediately followed by a single nonterminal symbol. Sample valid rules would then include

$$X \to y$$
$$X \to yZ.$$

However, a rule such as

$$A \to Ab$$

is not allowed.

Context Free (Type 2) Grammars

Unlike the type 3 grammar, which limits the right hand side of a rule to a terminal symbol, optionally followed by a non-terminal symbol, a *type 2* or *context free* grammar allows any combination of non-terminal and terminal symbols on the right hand side of a production rule. The right hand side may also be null. The left hand side of a type 2 rule is always a single, non-terminal symbol. Thus, in a rule of the form $\alpha C \beta \rightarrow \alpha d \beta$, both α and β must be null.

Context free production rules might look like:

$$A \rightarrow Ab$$
$$A \rightarrow AB$$
$$A \rightarrow \phi$$

Context Sensitive (Type 1) Grammars

A *context sensitive, context dependent,* or *Type 1* grammar allows for rules such as $\alpha A \beta \rightarrow \alpha \psi \beta$ where α and β are possibly null sequences. Because there can be a "context" allowed on the left hand side of a rule, it is referred to as context-sensitive grammar. The variable ψ represents a non-empty sequence of symbols, removing the capability of a production rule to delete a symbol. The length of the left hand side of this rewrite rule is always less than or equal to the length of the RHS.

Context free languages that don't have the null string are a proper subset of the context sensitive languages. The following thus are valid rules in a context sensitive grammar

$$RstU \rightarrow Rstu$$
$$R \rightarrow Rs$$

while the following is not valid

$$Rt \rightarrow t.$$

Phrase Structure (Type 0) Grammars

The most general type of grammar is a phrase structure grammar, which allows the context sensitive capability of a type 1 grammar and allows the null-string to be produced in the language.

Grammar Types and Natural Language

While any of these types of grammars could be the basis for most natural languages, context free grammars appear to be satisfactory approximations of a human's grammar in most situations. Partee, Meulen and Wall (Partee, Meulen, & Wall, 1990) note that "it took nearly thirty years to find one completely convincing example of a natural language which was not context free." Adding some of the features of a context sensitive grammar are necessary in some cases, and a number of mildly context sensitive grammars have been proposed, including index grammars and head grammars, that may prove useful when limiting the application of (otherwise context-free) rules to specific terms or situations.

Because context-free grammars are adequate in many situations and are mathematically far more tractable than are other grammars, we will examine the probabilities of rules assuming that the grammar is consistent with the assumptions of a context free grammar.

The Probability of a Parse

It is easy to compute the probability of a fully tagged statement, produced by a parser consistent with a context free grammar (Charniak, 1993; Fu, 1982; Lee & Fu, 1972). A fully tagged and parsed sentence is the original set of terms along with the full parse tree. This is the set of all tags (non-terminal symbols) used in the parse and terms (terminal symbols) from the statement. Computing the probability of a parse given that it is in the set of relevant documents or the probability of a parse from the set of all documents allows us to use the analytic methods to compute the expected performance for a retrieval system with documents written in a natural language.

Each production rule has an assigned probability, computed or estimated (Fu, 1982) so that the sum of probabilities associated with the set of rules having a given left hand side x, such as

$$x \to a$$
$$x \to b$$
$$x \to \ldots$$
$$x \to n,$$

is 1. The probability of a given parse is the product of the probability of each of the rules that is applied in producing the full parse. If the parsings of different statements are statistically independent, then the probability of the entire parse may be computed by multiplying the probabilities of the independent parses. Text filtering and retrieval performance may be predicted using techniques incorporating term dependencies described in Chapter 8. Computing the performance with these methods, we may make a number of performance claims about systems incorporating natural language capabilities.

EXERCISE 10.9. [*Moderate*] Develop a set of syntactic rules for the following arithmetic statements: $3 + 2 = 5$, $12 - 4 = 8$, $3 + 3 + 4 = 10$, and $3 = 3$. What is the probability for each rule given this data set?

Table 10.2. Sample sentences, part-of-speech tags and relevance values for 10 statements. Probabilities are for applied rewrite rules for the sentences. LHS represents the left hand side of a rewrite rule, and RHS represents the right hand side of a rule.

Sentence	POS Tags	Rel.	Sentence	POS Tags	Rel.
v w	A B	R	x w	A B	R
y w	A B	R	v z	A B	N
v y	A C	N	w v	B A	N
w z	B C	N	v w	A C	N
y w	A B	N	w x	C B	N

Rule	Pr($RHS|LHS$)	Pr($RHS|LHS$, Rel)
$S \to RL$	1	1
$L \to A$	7/10	3/3
$L \to B$	2/10	0
$L \to C$	1/10	0
$R \to A$	1/10	0
$R \to B$	6/10	3/3
$R \to C$	3/10	0
$A \to v$	5/8	1/3
$A \to x$	1/8	1/3
$A \to y$	2/8	1/3
$B \to w$	6/8	3/3
$B \to x$	1/8	0
$B \to z$	1/8	0
$C \to w$	2/4	—
$C \to y$	1/4	—
$C \to z$	1/4	—

10.7   PERFORMANCE WITH MULTIPLE SYNTACTIC TAGS: A CASE STUDY

Term Independence Performance

Consider a query that consists of only the term w. Using the data in Table 10.2, we find that among the ten sentences, there are 8 sentences with the query term w (so $t = .8$) and from among the 3 relevant sentences all have w (thus $p = 1$). Thus, using $\mathcal{A} = (1 - p + t)/2$, $\mathcal{A} = (1 - 1 + .8)/2 = .4$.

Using Syntactic Structures

Consider a query for w tagged so that it as at the end of the sentence, in the rightmost position, e.g. $R - w$, with optionally any preceding non-terminal symbol. Here R represents "right." Looking for w in sentence final position, we find that this occurs five times in the ten sentences, and thus $t = .5$, while among the relevant sentences, it occurs in all the relevant sentences. We thus find that $A = (1 - 1 + .5)/2 = .25$. Using linguistic tagging in this case provides a marked improvement over the results obtained with the binary independent model, from $A = .4$ improving to $A = .25$.

Now, let us assume that the query is further refined so that it is $R - B - w$, that is, w of part-of-speech B of part-of-speech R. The probability that we achieve this from among the set of all sentences is $t = \Pr(R - B - w) = 4/10$. Using similar techniques, $\Pr(R - B - w|rel) = 1$. The optimal performance may thus be computed as $A = (1 - 1 + 4/10)/2 = .2$, an improvement over the previous case.

Clearly, the more grammatical knowledge one has of a term in this example, the better will be the filtering performance.

10.8 EVALUATING GRAMMAR QUALITY WITH RETRIEVAL PERFORMANCE·

Evaluating one or more grammars that might be used with a filtering system can allow one to select a good, or possibly the best grammar, for a particular application or family of applications. We can describe the performance relationship between grammars by noting the ASL values obtained with two grammars for either one query, or the ASL for two grammars taken over a range of queries. Some grammars are expected to be better than others for all cases where all query terms are positive or neutral discriminators. For example, when grammar $\mathcal{G}_i$ can discriminate between terms that a second grammar $\mathcal{G}_j$ can't, and the grammars are otherwise equivalent, then $\mathcal{G}_i$ is expected to be superior to $\mathcal{G}_j$.

Definition 10.2 (Grammar Superiority) *Given the ASL of grammar* $\mathcal{G}$, *denoted as* $ASL(\mathcal{G})$, *we denote the superiority or equality of grammar i over grammar j as*

$$\mathcal{G}_{i,x} \overset{*}{\geq} \mathcal{G}_{j,x}$$

for domain x when $ASL(\mathcal{G}_{i,x}) \leq ASL(\mathcal{G}_{j,x})$.

Theorem 10.7 (Transitivity of grammar superiority) *If* $\mathcal{G}_{i,x} \overset{*}{\geq} \mathcal{G}_{j,x}$ *and* $\mathcal{G}_{j,x} \overset{*}{\geq} \mathcal{G}_{k,x}$ *then* $\mathcal{G}_{i,x} \overset{*}{\geq} \mathcal{G}_{k,x}$ *for all i, j, k, and x.*

Informally, we may describe one grammar as *superior* to another grammar for a given query if the A values for the performatively superior grammar are less than the A values for the inferior grammar.   A *performance hierarchy* of grammars is as follows:

1. **Superiority occurs in a domain**

   $\mathcal{G}_i$ is *performatively superior* to (or equal to) $\mathcal{G}_j$ in domain x when $\mathcal{G}_{i,x} \overset{*}{\geq} \mathcal{G}_{j,x}$ with the expected ASL for the grammars taken over domain x.

2. **Superiority occurs in all cases studied**

   $\mathcal{G}_i$ is *universally superior* to (or equal to) $\mathcal{G}_j$ when $\mathcal{G}_i$ is *performatively superior* to (or equal to) $\mathcal{G}_j$ in domain x, and x is the universe of all extant domains. This is an empirical superiority.

3. **Superiority is necessary**

   $\mathcal{G}_i$ is *necessarily superior* to $\mathcal{G}_j$ when it is necessary that $\mathcal{G}_{i,x} \overset{*}{\geq} \mathcal{G}_{j,x}$ for all possible domains x.

Consider the case where there are a set of queries, some with *run* as a noun and some with *run* as a verb. Assume that we have a grammar $\mathcal{G}_0$ which has no part-of-speech tagging, essentially a null grammar. We also have $\mathcal{G}_{nv}$ which can perfectly parse all sentences and tag *nouns* and *verbs* as such. In some situations, queries will specify that in the case of ambiguous terms, the verb form is the one that is desired or the nominal form is the one that is sought. This would lead to improved performance for many of the queries that can use the partial-disambiguation that is available through use of parts-of-speech. One thus expects that $\mathcal{G}_{nv} \overset{*}{\geq} \mathcal{G}_0$ and we can say that $\mathcal{G}_{nv}$ is necessarily superior to $\mathcal{G}_0$ if query terms are positive discriminators.

Further, if we have a grammar $\mathcal{G}_{n_{123}v}$ that can distinguish between verbs and nouns and can also distinguish three different kinds of nouns, and these noun types occur in some queries, we would expect $\mathcal{G}_{n_{123}v}$ to be necessarily superior to $\mathcal{G}_{nv}$ when pairs of query terms and part-of-speech tags are positive discriminators.

A weaker case, where only performative superiority would hold, would be a situation where a grammatical tagger that only tagged nouns (leaving all other terms as being the same) was shown to produce better retrieval on an average than a tagger that only tagged adjectives (leaving all other terms as being the same). The performative difference between these two is an empirical question (at present) and we than thus say that the noun tagging grammar $\mathcal{G}_n$ is only performatively superior to the adjective tagging grammar $\mathcal{G}a$.

Note that necessary superiority, or conditions under which necessary superiority holds, may be determined using analytic methods.

Language Supporting Actions

Our ability to analytically study the retrieval and filtering of natural language statements can lead us to a more general model of how natural languages allow us to discriminate between linguistic expressions (messages) that facilitate human communication and decision making. Consider a message M of benefit b to a human. We assume that messages having benefit b or over are considered relevant and messages with benefit below b to be non-relevant. We assume that a particular benefit is due to an association between the message and a physical action that may be taken with associated benefit b. This message has the average position of $\mathcal{A}$.

Assume that we have a set of part-of-speech tags T for the language and a set of syntactic rules $\mathcal{R}$. The grammar $\mathcal{G}$ is a function described earlier in the chapter. Let us also assume that there exists an error function $\mathcal{H}$ that is due to the human constraints on the amount and rate at which humans can learn and currently operate. The $\mathcal{H}$ function is a source of error, as is a Q less than 1. This $\mathcal{H}$ function is assumed to be non-linearly related to the number and complexity of syntactic rules $\mathcal{R}$ and tags T. The set of tags T and the assignments of tags to message terms is such that a certain level of filtering performance, $\mathcal{A}$, is obtained.

Conjecture 10.1 (Optimal language) *An optimal language, given a world where benefits b are associated with receiving messages M, is such that parameters $\mathcal{R}$ and T minimize*

$$E\left(ASL\left(A\big(\mathcal{G}(T, \mathcal{R})\big), \mathcal{H}(T, \mathcal{R}) \right), M \right) \tag{10.10}$$

over the set of messages encountered.

Consider what happens when the number of part-of-speech (POS) tags and syntactic rules increases. As the number of part-of-speech tags increase, $\mathcal{A}$ will decrease (improve) as the ability to discriminate between potentially ambiguous terms is increased. However, performance will have a component that tends toward decreasing performance as the error function returns a greater error value that is proportional to the increased number of part-of-speech tags, system complexity, and the increased chance for human error due to the difficulty associated with learning a more complex language.

A human system that must discriminate and make decisions based on natural language will discriminate best when using an optimal language. We may assume that an optimal language is achieved when the ASL value in Equation 10.10 is minimized. This is obtained at a given syntactic complexity and a given number of part-of-speech tags and represents the break-even point between these two conflicting factors. At this point, the performance decrease in error rate due to increased language complexity is the same as the performance increase due to the increased value of discrimination capability that is available with increasingly complex linguistic structures.

The error function for humans may be learned if natural language is assumed to be optimized for different applications and sublanguages, e.g. children vs. adults, folksy chatting vs. technical reports, etc. (Sager, 1981a). By learning the functional values for several different language types, the error function may be estimated and its general characteristics understood.

10.9   DISCUSSION

The difficulties associated with satisfactorily incorporating natural language processing into retrieval systems present unique challenges to those estimating filtering and retrieval performance. Part-of-speech tags can help filtering systems separate relevant documents from non-relevant documents. A rule was proposed (Equation 10.1) using these tags that suggested that improvement would occur for retrieving individual terms for certain ranges of parameter values associated with the tagging process. Tagging with multiple terms and the stemming of suffixes similarly may improve performance.

The relationships between terms may improve filtering system performance. Grammars may be used to parse statements, providing part-of-speech tags and implicitly providing information about the relationships between terms. Methods described in Chapter 7 can use these term relationships to provide improved document ranking. The case study provided in this chapter shows how increasing the amount of linguistically provided information can improve performance.

Bibliography

Allan, J. (1995). Relevance feedback with too much data. In *ACM Annual Conference on Research and Development in Information Retrieval*, pp. 337–343 New York. ACM Press.

Allan, J. (1996). Incremental relevance feedback for infomation filtering. In *Proceedings of the 19th Annual International ACM SIGIR Conference on Research and Development in Information Retrieval, Zurich, Switzerland*, pp. 270–278 New York. ACM Press.

Anderson, A. R., & Belnap, N. D. (1975). *Entailment: The Logic of Relevance and Necessity*. Princeton University Press.

Bahadur, R. R. (1961). A representation of the joint distribution of response of n dichotomous items. In Solomon, H. (Ed.), *Studies in Item Analysis and Prediction*, pp. 158–168. Stanford U. Press, Stanford, CA.

Baker, S. L. (1986). Overload, browsers, and selections. *Library and Information Science Research, 8*(4), 315–329.

Barry, C. L. (1994). User-defined relevance criteria: An exploratory study. *Journal of the American Society for Information Science, 45*(3), 149–159.

Beckman, F. S. (1980). *Mathematical Foundations of Programming*. Addison Wesley, Reading, Mass.

Belkin, N. J., & Croft, W. B. (1992). Information filtering and information retrieval: Two sides of the same coin. *Communications of the ACM, 35*(12), 29–38.

Berger, A. L., Della Pietra, V. J., & Della Pietra, S. A. (1996). A maximum entropy approach to natural language processing. *Computational Linguistics, 22*(1), 39–71.

Bishop, C. M. (1995). *Neural Networks for Pattern Recognition*. Oxford University Press, New York.

Blair, D. C., & Maron, M. E. (1985). An evaluation of retrieval effectiveness for a full-text document-retrieval system. *Communications of the ACM, 28*(3), 289–299.

Boll, J. J. (1985). *Shelf Browsing, Open Access and Storage Capacity in Research Libraries*. No. 169 in Occasional Papers. U. of Illinois, Graduate School of Library Science, Urbana, IL.

Bollmann, P., & Cherniavsky, V. (1981). Measurement-theoretical investigation of the MZ-metric. In Oddy, R., Robertson, S. E., van Rijsbergen, C. J., & Williams, P. W. (Eds.), *Information Retrieval Research*, pp. 256–267 London. Butterworths.

Bookstein, A. (1979). Relevance. *Journal of the American Society for Information Science, 30*(5), 269–273.

Bookstein, A. (1982). Explanation and generalization of vector models in information retrieval. In *Research and Development in Information Retrieval: Proceedings of the*

5th International Conference on Information Retrieval, pp. 118–132 Berlin. Springer-Verlag.

Bookstein, A. (1983). Information retrieval: A sequential learning process. *Journal of the American Society for Information Science, 34*(4), 331–342.

Bookstein, A. (1985). Implications of Boolean structure for probabilistic retrieval. In *Proceedings of the Eigth Annual International ACM SIGIR Conference on Research and Development in Information Retrieval*, pp. 11–17. ACM Press.

Bovey, J. D., & Robertson, S. E. (1984). An algorithm for weighted searching on a Boolean system. *Information Technology, 3*(2), 84–87.

Boyce, B. R., & McLain, J. P. (1989). Entry point depth and online search using a controlled vocabulary. *Journal of the American Society for Information Science, 40*(4), 273–276.

Brewka, B. (1991). *Nonmonotonic Reasoning: Logical Foundations of Commonsense.* Cambridge University Press, Cambridge, UK.

Brill, E. (1994). Some advances in transformation-based part of speech tagging. In *Proceedings of the Twelfth National Conference on Artificial Intelligence (AAAI-94)*, pp. 722–727 Menlo Park, CA. AAAI Press.

Brookes, B. (1968). The measures of information retrieval effectiveness proposed by Swets. *Journal of Documentation, 24*(1), 41–54.

Buckles, B. P., & Petry, F. E. (1982). A fuzzy representation of data for relational databases. *Fuzzy Sets and Systems, 7*(3), 213–226.

Buckles, B. P., & Petry, F. E. (1983). Information theoretical characterization of fuzzy relational databases. *IEEE Transactions on Systems, Man, and Cybernetics, SMC-13*(1), 74–77.

Burgin, R., & Dillon, M. (1992). Improving disambiguation in FASIT. *Journal of the American Society for Information Science, 43*, 101–114.

Callan, J. P. (1994). Passage-level evidence in document retrieval. In *Proceedings of the 17th Annual International ACM SIGIR Conference on Research and Development in Information Retrieval, Dublin, Ireland*, pp. 302–310 New York. ACM Press.

Callan, J. P., Lu, Z., & Croft, W. B. (1995). Searching distributed collections with inference nets. In *Proceedings of the 18th Annual International ACM SIGIR Conference on Research and Development in Information Retrieval, Seattle, Washington*, pp. 21–28 New York. ACM Press.

Charniak, E. (1993). *Statistical Language Learning.* MIT Press, Cambridge, Mass.

Charniak, E., & McDermott, D. (1985). *Introduction to Artificial Intelligence.* Addison-Wesley, Reading, Mass.

Chomsky, N. (1965). *Aspects of the Theory of Syntax.* MIT Press, Cambridge, Mass.

Chow, C., & Liu, C. (1968). Approximating discrete probability distributions with dependence trees. *IEEE Transactions on Information Theory, IT-14*(3), 462–467.

Church, K. W. (1995). One term or two. In *Proceedings of the 18th Annual International ACM SIGIR Conference on Research and Development in Information Retrieval, Seattle, Washington*, pp. 310–318 New York. ACM Press.

Clifford, H., & Stephenson, W. (1975). *An Introduction to Numerical Classification.* Academic Press, New York.

Conrad, J. G., & Utt, M. H. (1994). A system for discovering relationships by feature extraction from text dictionaries. In *Proceedings of the 17th Annual International ACM SIGIR Conference on Research and Development in Information Retrieval, Dublin, Ireland*, pp. 260–270 New York. ACM Press.

Conway, J. H., Sloane, N. J. A., & Wilks, A. R. (1989). Gray codes for reflection groups. *Graphs and Combinatorics, 5*, 315–325.

Cooper, W. S. (1968). Expected search length: A single measure of retrieval effectiveness based on weak ordering action of retrieval systems. *Journal of the American Society for Information Science, 19*(1), 30–41.

Cooper, W. S. (1973). On selecting a measure of retrieval effectiveness. *Journal of the American Society for Information Science, 24*, 87–100.

Cooper, W. S. (1994). The formalism of probability theory in IR: A foundation or an encumbrance?. In *Proceedings of the 17th Annual International ACM SIGIR Conference on Research and Development in Information Retrieval, Dublin, Ireland*, pp. 242–247 New York. ACM Press.

Cooper, W. S. (1995). Some inconsistencies and misidentified modeling assumptions in probabilistic information retrieval. *ACM Transactions on Information Systems, 13*(1), 100–111.

Cooper, W. S., & Huizinga, P. (1982). The maximum entropy principle and its application to the design of probabilistic retrieval systems. *Information Technology: Research and Development, 1*(2), 99–112.

Cover, J. F., & Walsh, B. C. (1988). Online text retrieval via browsing. *Information Processing and Management, 24*(1), 31–37.

Cox, D. R., & Wermuth, N. (1991). A simple approximation for bivariate and trivariate normal integrals. *International Statistical Review, 59*(2), 263–269.

Crestani, F., & van Rijsbergen, C. J. (1995). Probability kinematics in information retrieval. In *Proceedings of the 18th Annual International ACM SIGIR Conference on Research and Development in Information Retrieval, Seattle, Washington*, pp. 291–299 New York. ACM Press.

Croft, W. B. (1986). Boolean queries and term dependencies in probabilistic retrieval models. *Journal of the American Society for Information Science, 37*(2), 71–77.

Croft, W. B., & Harper, D. (1979). Using probabilistic models of document retrieval without relevance information. *Journal of Documentation, 35*(4), 285–295.

Deerwester, S., Dumais, S. T., Furnas, G. W., Landauer, T. K., & Harshman, R. (1990). Indexing by latent semantic analysis. *Journal of the American Society for Information Science, 41*(6), 391–407.

Diaz, M. R. (1981). *Topics in the Logic of Relevance.* Philosophia Verlag, Munich.

Dorfman, J. H. (1997). *Bayesian Economics Through Numerical Methods.* Springer-Verlag, New York.

Drezner, Z. (1992). Computation of the multivariate normal integral. *ACM Transactions on Mathematical Software, 18*(2), 470–480.

Dunlop, M. D. (1997). Time, relevance and interaction modelling for information retrieval. In *Proceedings of the 20th Annual International ACM SIGIR Conference on Research and Development in Information Retrieval, Philadelphia, Pennsylvania*, pp. 206–213. ACM Press, New York.

Edwards, A. W. F. (1972). *Likelihood.* Cambridge, Cambridge, England.

Efthimiadis, E. N. (1996). Query expansion. In *Annual Review of Information Science and Technology*, pp. 121–187. Information Today, Inc., Medford, NJ.

Egan, J. P. (1975). *Signal Detection Theory and ROC Analysis.* Academic Press, New York.

Eisenberg, M. B. (1988). Measuring relevance judgments. *Information Processing and Management, 24*(3), 373–389.

Evans, R. (1994). Beyond Boolean: Relevance ranking, natural language and the new search paradigm. In *Proceedings of the Fifteenth National Online Meeting*, pp. 121–128 Medford, NJ. Learned Information.

Everett, D. M., & Carter, S. C. (1992). Topology of document retrieval systems. *Journal of the American Society for Information Science, 43*(10), 658–673.

Feys, R. (1965). *Modal Logics*. E. Nauwelaerts, Louvain.

Fidel, R. (1992). Who needs controlled vocabulary?. *Special Libraries, 83*, 1–9.

Fidel, R., & Crandall, M. (1997). Users' perception of the performance of a filtering system. In *Proceedings of the 20th Annual International ACM SIGIR Conference on Research and Development in Information Retrieval, Philadelphia, Pennsylvania*, pp. 198–205. ACM Press, New York.

Fillmore, C. (1977). The case for case. In Bach, E., & Harms, R. (Eds.), *Universals in Linguistic Theory*, pp. 1–88. Holt, Rinehart, and Winston, New York.

Flores, I. (1956). Reflected number systems. *IRE Transactions on Electronic Computers, EC-5*(2), 79–82.

Fox, E. A., & Winett, S. G. (1990). Using vector and extended Boolean matching in an expert system for selecting foster homes. *Journal of the American Society for Information Science, 41*(1), 10–26.

Fu, K. S. (1982). *Syntactic Pattern Recognition*. Prentice-Hall, Englewood Cliffs, NJ.

Fuhr, N. (1989). Optimum polynomial retrieval functions. In *ACM Annual Conference on Research and Development in Information Retrieval*, pp. 69–76 New York. ACM Press.

Gärdenfors, P. (1992). Belief revision: An introduction. In Gärdenfors, P. (Ed.), *Belief Revision*, pp. 1–28. Cambridge University Press, Cambridge.

Gelman, A., Carlin, J. B., Stern, H. S., & Rubin, D. B. (1995). *Bayesian Data Analysis*. Chapman and Hall, London.

Geman, S., & Geman, D. (1984). Stochastic relaxation, Gibbs distribution, and the Bayesian restoration of images. *IEEE Transactions on Pattern Analysis and Machine Intelligence, 6*, 721–741.

Genesereth, M. R., & Nilsson, N. J. (1987). *Logical Foundations of Artificial Intelligence*. Morgan Kaufmann, Los Altos.

Gensler, H. J. (1990). *Symbolic Logic: Classical and Advanced Systems*. Prentice Hall, Englewood Cliffs, N.J.

Gey, F. C. (1993). *Probabilistic Dependence and Logistic Inference in Information Retrieval*. Ph.D. thesis, U. of California, Berkeley.

Gilbert, E. N. (1958). Gray codes and paths on the n-cube. *Bell System Technical Journal, 37*, 815–826.

Gordon, M. D., & Lenk, P. (1992). When is the probability ranking principle suboptimal?. *Journal of the American Society for Information Science, 43*(1), 1–14.

Greiff, W. R., Croft, W. B., & Turtle, H. (1997). Computationally tractable probabilistic modeling of Boolean operators. In *Proceedings of the 20th Annual International ACM SIGIR Conference on Research and Development in Information Retrieval, Philadelphia, Pennsylvania*, pp. 119–128. ACM Press, New York.

Groenink, A. (1997). *Surface Without Structure*. Ph.D. thesis, University of Utrecht, Utrecht, Netherlands.

Haas, S. W., & Losee, R. M. (1994). Looking in text windows: Their size and composition. *Information Processing and Management, 30*(5), 619–629.

Haight, F. A. (1967). *Handbook of the Poisson Distribution*. John Wiley, New York.

Hamming, R. (1986). *Coding and Information Theory* (Second edition). Prentice-Hall, Englewood Cliffs, N.J.

Harman, D. W. (1991). How effective is suffixing. *Journal of the American Society for Information Science, 42*(1), 7–15.

Harter, S. P. (1975a). Probabilistic approaches to automatic keyword indexing: Part I. *Journal of the American Society for Information Science, 26*(4), 197–206.

Harter, S. P. (1975b). Probabilistic approaches to automatic keyword indexing: Part II. an algorithm for probabilistic indexing. *Journal of the American Society for Information Science, 26*(5), 280–289.

Heckerman, D., Geiger, D., & Chickering, D. M. (1995). Learning Bayesian networks: The combination of knowledge and statistical data. *Machine Learning, 20*(3), 197–243.

Heine, M. H. (1973). Distance between data as an objecvtive measure of retrieval effectiveness. *Information Processing and Management, 9,* 181–198.

Huestis, J. C. (1988). Clustering LC classification numbers in an online catalog for improved browsability. *Information Technology and Libraries, 7*(4), 381–393.

Hughes, G. E., & Cresswell, M. J. (1968). *An Introduction to Modal Logic.* Methuen, London.

Hull, D. (1994). Improving text retrieval for the routing problem using latent semantic indexing. In *Proceedings of the 17th Annual International ACM SIGIR Conference on Research and Development in Information Retrieval, Dublin, Ireland,* pp. 282–291 New York. ACM Press.

Hull, D. A., Pedersen, J. O., & Schutze, H. (1996). Method combination for document filtering. In *Proceedings of the 19th Annual International ACM SIGIR Conference on Research and Development in Information Retrieval, Zurich, Switzerland,* pp. 279–287. ACM Press, New York.

Hurt, C. D. (1998). Nonmonotonic logic for use in information retrieval: An exploratory paper. *Information Processing and Management, 34*(1), 35–41.

Hwang, P., & Burgers, W. P. (1997). Properties of trust: An analytical view. *Organizational Behavior and Human Decision Processes, 69*(1), 67–73.

Iivonen, M., & Sonnenwald, D. H. (1998). From translation to navigation of different discourses: A model of search term selection during the pre-online stage of the search process. *Journal of the American Society for Information Science, 49*(4), 312–326.

Jacquemin, C. (1996). What is the tree that we see through the window: a linguistic approach to windowing and term variation. *Information Processing and Management, 32,* 445 448.

Johnson, N. L., & Kotz, S. (1972). *Distributions in Statistics: Continuous Multivariate Distributions.* John Wiley & Sons, Inc., New York.

Johnson, N. L., Kotz, S., & Balakrishnan, N. (1997). *Discrete Multivariate Distributions.* John Wiley & Sons, Inc., New York.

Johnson, W. O., & Kokolakis, G. E. (1994). Bayesian classification based on multivariate binary data. *J. of Statistical Planning and Inference, 41,* 21–35.

Kantor, P. B. (1984). Maximum entropy and the optimal design of automated information retrieval systems. *Information Technology: Research and Development, 3*(2), 88–94.

Kneale, W., & Kneale, M. (1984). *The Development of Logic.* Clarendon Press, Oxford.

Knill, K., & Young, S. (1997). Hidden Markov models in speech and language processing. In Young, S., & Bloothooft, G. (Eds.), *Corpus-Based Methods in Language and Speech Processing,* pp. 27–68. Kluwer, Dordrecht.

Kraaij, W., & Pohlmann, R. (1996). Viewing stemming as recall enhancement. In *Proceedings of the 19th Annual International ACM SIGIR Conference on Research and Development in Information Retrieval, Zurich, Switzerland,* pp. 40–48 New York. ACM Press.

Kraft, D. H., & Bookstein, A. (1978). Evaluation of information retrieval systems: A decision theory approach. *Journal of the American Society for Information Science, 29*(1), 31–40.

Kraft, D. H., & Lee, T. (1979). Stopping rules and their effect on expected search length. *Information Processing and Management, 15*(1), 47–58.

Kraft, D. H., & Waller, W. (1981). A Bayesian approach to user stopping rules for information retrieval systems. *Information Processing and Management, 17*(6), 349–361.

Krovetz, R., & Croft, W. B. (1992). Lexical ambiguity and information retrieval. *ACM Transactions on Information Systems, 10*, 115–141.

Lalmas, M. (1998). Logical models in information retrieval: Introduction and overview. *Information Processing and Management, 34*(1), 19–33.

Lam, W., Mukhopadhyay, S., Mostafa, J., & Palakal, M. (1996). Detection of shifts in user interests for personalized information filtering. In *Proceedings of the 19th Annual International ACM SIGIR Conference on Research and Development in Information Retrieval, Zurich, Switzerland*, pp. 317–325 New York. ACM Press.

Lee, H. C., & Fu, K. S. (1972). A stochastic syntax analysis procedure and its application to pattern classification. *IEEE Transactions on Computers, C-21*(7), 660–666.

Lee, J. H. (1994). Properties of extended Boolean models in information retrieval. In *ACM Annual Conference on Research and Development in Information Retrieval*, pp. 182–190 New York. ACM Press.

Lee, P. M. (1989). *Bayesian Statistics: An Introduction.* Oxford University Press, New York.

Lee, T. T. (1987). An information theoretic analysis of relational databases, parts I and II. *IEEE Transactions on Software Engineering, SE-13*(10), 1049–1072.

Lenk, P. J. (1990). Bayesian predictive distributions under multinomial sampling. In Geisser, S., Hodges, J. S., Press, S. J., & Zellner, A. (Eds.), *Bayesian and Likelihood Methods in Statistics and Econometrics*, pp. 357–370. North-Holland, Amsterdam.

Levenstein, V. I. (1966). Binary codes capable of correcting deletions, insertions and reversals. *Cybernet. Control Theor., 10*, 707–710.

Lewis, D. D., & Sparck-Jones, K. (1996). Natural language processing for information retrieval. *Communications of the ACM, 39*(1), 92–101.

Littman, M. L., & Jiang, F. (1998). A comparison of two corpus-based methods for translingual information retrieval. CS Department, Duke University.

Losee, R. M. (1987). Probabilistic retrieval and coordination level matching. *Journal of the American Society for Information Science, 38*(4), 239–244.

Losee, R. M. (1988). Parameter estimation for probabilistic document retrieval models. *Journal of the American Society for Information Science, 39*(1), 8–16.

Losee, R. M. (1989). Minimizing information overload: The ranking of electronic messages. *Journal of Information Science, 15*(3), 179–189.

Losee, R. M. (1990). *The Science of Information: Measurement and Applications.* Academic Press, San Diego.

Losee, R. M. (1991). An analytic measure predicting information retrieval system performance. *Information Processing and Management, 27*(1), 1–13.

Losee, R. M. (1993a). The relative shelf location of circulated books: A study of classification, users, and browsing. *Library Resources & Technical Services, 37*(2), 197–209.

Losee, R. M. (1993b). Seven fundamental questions for the science of library classification. *Knowledge Organization, 20*(2), 65–70.

Losee, R. M. (1994a). Term dependence: Truncating the Bahadur Lazarsfeld expansion. *Information Processing and Management, 30*(2), 293–303.

Losee, R. M. (1994b). Upper bounds for retrieval performance and their use measuring performance and generating optimal Boolean queries: Can it get any better than this?. *Information Processing and Management, 30*(2), 193–203.

Losee, R. M. (1996). Text windows and phrases differing by discipline, location in document, and syntactic structure. *Information Processing and Management, 32*(6), 747–767.

Losee, R. M. (1997a). Browsing document collections: Automatically organizing digital libraries and hypermedia using the Gray code. *Information Processing and Management, 33*(2), 175–192.

Losee, R. M. (1997b). Comparing Boolean and probabilistic information retrieval systems across queries and disciplines. *Journal of the American Society for Information Science*, *48*(2), 143–156.

Losee, R. M. (1997c). A discipline independent definition of information. *Journal of the American Society for Information Science*, *48*(3), 254–269.

Losee, R. M., & Bookstein, A. (1988). Integrating Boolean queries in conjunctive normal form with probabilistic retrieval models. *Information Processing and Management*, *24*(3), 315–321.

Losee, R. M., Bookstein, A., & Yu, C. T. (1986). Probabilistic models for document retrieval: A comparison of performance on experimental and synthetic databases. In *ACM Annual Conference on Research and Development in Information Retrieval*, pp. 258–264.

MacKay, D. J. C. (1996). Equivalence of linear Boltzmann chains and hidden Markov models. *Neural Computation*, *8*(1), 178–181.

Mandelbrot, B. (1961). On the theory of word frequencies and on related Markovian models of discourse. In *Structure of Language and Its Mathematical Aspects: Proceedings of Symposia in Applied Mathematics, vol. XII*, pp. 190–219. American Mathematical Society.

Marchionini, G. (1987). An invitation to browse. *Canadian Journal of Information Science*, *12*(3/4), 69–79.

Margulis, E. L. (1993). Modelling documents with multiple Poisson distributions. *Information Processing and Management*, *29*, 215–227.

Maritz, J. S. (1970). *Empirical Bayes Methods*. Methuen, London.

Markowitz, J. A. (1977). *A Look at Fuzzy Categories*. Ph.D. thesis, Northwestern University.

Maron, M. E., & Kuhns, J. L. (1960). On relevance, probabilistic indexing, and information retrieval. *Journal of the ACM*, *7*, 216–244.

Martin, M. C. (1995). Situating relevance in information retrieval: Every small step counts. Master's thesis, U. of North Carolina, Chapel Hill, NC.

McCarthy, J. (1980). Circumspection—a form of non-monotonic reasoning. *Artificial Intelligence*, *13*, 27–39.

Metropolis, N., & Ulam, S. (1949). The Monte Carlo method. *Journal of the American Statistical Association*, *44*, 335–341.

Metzing, D. (Ed.). (1979). *Frame Conceptions and Text Understanding*. Walter De Gruyter and Co., Berlin.

Millman, R. S., & Parker, G. D. (1981). *Geometry: A Metric Approach with Models*. Springer-Verlag, New York.

Moore, R. C. (1985). Semantical considerations on nonmonotonic logic. *Artificial Intelligence*, *25*(1), 75–94.

Morse, P. M. (1970). Search theory and browsing. *Library Quarterly*, *40*(4), 391–408.

Myers, E. W., & Miller, W. (1988). Optimal alignments in linear space. *Computer Applications in Biosciences*, *4*, 11–17.

Negoita, C. V. (1973). On the notion of relevance in information retrieval. *Kybernetes*, *2*, 161–165.

Ney, H. (1997). Corpus-based statistical methods in speech and language processing. In Young, S., & Bloothooft, G. (Eds.), *Corpus-Based Methods in Language and Speech Processing*, pp. 1–26. Kluwer, Dordrecht.

Ng, K.-C., & Abramson, B. (1990). Uncertainty management in expert systems. *IEEE Expert—Intelligent Systems & Their Applications*, *5*(2), 29–48.

Nie, J. (1989). An information retrieval model based on modal logic. *Information Processing and Management, 25*(5), 477–491.

Nilsson, N. J. (1990). Probabilistic logic. In Shafer, G., & Pearl, J. (Eds.), *Readings in Uncertain Reasoning*, pp. 680–688. Morgan Kaufmann, San Mateo, CA.

O Ruanaidh, J. J. K., & Fitzgerald, W. J. (1996). *Numerical Bayesian Methods Applied to Signal Processing*. Springer-Verlag, New York.

Paice, C. D. (1994). An evaluation method for stemming algorithms. In *Proceedings of the 17th Annual International ACM SIGIR Conference on Research and Development in Information Retrieval, Dublin, Ireland*, pp. 42–50 New York. ACM Press.

Pancha, P., & El Zarki, M. (1994). MPEG coding for variable bit rate video transmission. *IEEE Communication, 32*(5), 54–66.

Partee, B. H., Meulen, A. t., & Wall, R. E. (1990). *Mathematical Methods in Linguistics*. Kluwer, Dordrecht, The Netherlands.

Pitts, C. G. C. (1972). *Introduction to Metric Spaces*. Oliver and Boyd, Edinburgh.

Polson, N. (1996). Convergence of Markov Chain Monte Carlo algorithms. In *Proceedings of the Fifth Valencia International Conference on Bayesian Statistics*, pp. 297–322. Oxford University Press, New York.

Popovic, M., & Willett, P. (1992). The effectiveness of stemming for natural-language access to Slovene textual data. *Journal of the American Society for Information Science, 43*(5), 384–390.

Porter, M. F. (1980). An algorithm for suffix stripping. *Program, 14*(3), 130–137.

Pratt, W., Raiffa, H., & Schlaifer, R. (1995). *Introduction to Statistical Decision Theory*. MIT Press, Cambridge, Mass.

Press, S. J. (1972). *Applied Multivariate Analysis*. Holt, New York.

Prior, A. N. (1967). Modal logic. In Edwards, P. (Ed.), *The Encyclopedia of Philosophy*, Vol. 5, pp. 5–12. Macmillan, New York.

Przymusinski, T. C. (1990). Non-monotonic reasonoing versus logic programming: a new perspective. In Partridge, D., & Wilks, Y. (Eds.), *The Foundations of Artificial Intelligence*, pp. 49–71. Cambridge U. Press.

Radecki, T. (1979). Fuzzy set theoretical approach to document retrieval. *Information Processing and Management, 15*, 247–259.

Radecki, T. (1982). A probabilistic approach to information retrieval in systems with Boolean search request formulations. *Journal of the American Society for Information Science, 33*(6), 365–370.

Raghavan, V. V., Shi, H.-P., & Yu, C. T. (1983). Evaluation of the 2-Poisson model as a basis for using term frequency data in searching. In *ACM Annual Conference on Research and Development in Information Retrieval*, pp. 88–100 New York. ACM Press.

Ramsay, A. (1988). *Formal Methods in Artificial Intelligence*. Cambridge U. Press.

Reiter, R. (1980). Equality and domain closure in first order logic data bases. *Journal of the ACM, 27*, 235–249.

Reiter, R. (1990). Nonmonotonic reasoning. In Shafer, G., & Pearl, J. (Eds.), *Readings in Uncertain Reasoning*, pp. 637–656. Morgan Kaufmann, San Mateo, CA.

Resnick, P., & Varian, H. R. (1997). Recommender systems. *Communications of the ACM, 40*(3), 56–58.

Ribeiro, B. A. N., & Muntz, R. (1996). A belief network model for IR. In *Proceedings of the 19th Annual International ACM SIGIR Conference on Research and Development in Information Retrieval, Zurich, Switzerland*, pp. 253–260 New York. ACM Press.

Robertson, S. E., & Thompson, C. L. (1990). Weighted searching: the CIRT experiment. In *Informatics 10: Prospects for Intelligent Retrieval*, pp. 153–166. ASLIB, London.

Robertson, S. E., & Walker, S. (1994). Some simple effective approximations to the 2-Poisson model for probabilistic weighted retrieval. In *Proceedings of the 17th Annual International ACM SIGIR Conference on Research and Development in Information Retrieval, Dublin, Ireland*, pp. 232–241 New York. ACM Press.

Robertson, S. E. (1977). The probability ranking principle in IR. *Journal of Documentation, 33*(4), 294–304.

Robertson, S. E., & Sparck Jones, K. (1976). Relevance weighting of search terms. *Journal of the American Society for Information Science, 27*, 129–146.

Robertson, S. E., Van Rijsbergen, C. J., & Porter, M. (1981). Probabilistic models of indexing and searching. In Oddy, R., Robertson, S. E., van Rijsbergen, C. J., & Williams, P. W. (Eds.), *Information Retrieval Research*, pp. 35–56 London. Butterworths.

Rowe, N. C. (1988). *Artificial Intelligence Through Prolog*. Prentice Hall, Englewood Cliffs, N.J.

Russell, B. (1906). The theory of implication. *American Journal of Mathematics, 28*, 159–202.

Sager, N. (1981a). Information structures in texts of a sublangauge. In *Proceedings of the 44th ASIS Annual Meeting*, pp. 199–201 White Plains, NY. Knowledge Industry Publications.

Sager, N. (1981b). *Natural Language Information Processing: A Computer Grammar of English and Its Applications*. Addison-Wesley, Reading, Mass.

Sainsbury, M. (1991). *Logical Forms: An Introduction to Philosophical Logic*. Blackwell, Oxford, UK.

Salton, G., Wong, A., & Yu, C. T. (1976). Automatic indexing using term discrimination and term precision measurements. *Information Processing and Management, 12*(1), 43–51.

Salton, G. (1984). The use of extended Boolean logic in information retrieval. Tech. rep. TR 84–588, Cornell University, Computer Science Dept., Ithaca, N.Y.

Salton, G., & McGill, M. (1983). *Introduction to Modern Information Retrieval*. McGraw-Hill, New York.

Sanderson, M. (1994). Word sense disambiguation and information retrieval. In *Proceedings of the 17th Annual International ACM SIGIR Conference on Research and Development in Information Retrieval, Dublin, Ireland*, pp. 142–151 New York. ACM Press.

Sanford, D. H. (1989). *If P, then Q: Conditionals and the Foundations of Reasoning*. Routledge, London.

Saracevic, T. (1991). Individual differences in organizing, searching and retrieving information. In Griffiths, J. M. (Ed.), *Proceedings of the 54th ASIS Annual Meeting*, Vol. 28, pp. 82–86.

Saul, L. K., & Jordan, M. I. (1995). Boltzmann chains and hidden Markov models. In Tesauro, G., Touretzky, D. S., & Leen, T. K. (Eds.), *Advances in Neural Information Processing Systems*, Vol. 7, pp. 435–442. MIT Press, Cambridge, Mass.

Schamber, L. (1994). Relevance and information behavior. In Williams, M. E. (Ed.), *Annual Review of Information Science and Technology*, Vol. 29, pp. 3–48. American Society for Information Science, Washington, D.C.

Schamber, L., Eisenberg, M., & Nilan, M. S. (1990). A re-examination of relevance: Toward a dynamic, situational definition. *Information Processing and Management, 26*(6), 755–776.

Schutze, H., Hull, D. A., & Pedersen, J. O. (1995). A comparison of classifiers and document representations for the routing problem. In *Proceedings of the 18th Annual International ACM SIGIR Conference on Research and Development in Information Retrieval, Seattle, Washington*, pp. 229–237 New York. ACM Press.

Schweizer, B., & Sklar, A. (1963). Associative functions and abstract semigroups. *Publicationes Mathematicae Debrecen, 10*, 69–81.

Schweizer, B., & Sklar, A. (1983). *Probabilistic Metric Spaces*. North-Holland, New York.

Sebastiani, F. (1998). On the role of logic in information retrieval. *Information Processing and Management, 34*(1), 1–18.

Shaw, Jr., W. M. (1986). On the foundation of evaluation. *Journal of the American Society for Information Science, 37*(5), 346–348.

Shaw, Jr., W. M. (1995). Term-relevance computation and perfect retrieval performance. *Information Processing and Management, 31*(4), 491–498.

Shaw, Jr., W. M., Burgin, R., & Howell, P. (1997). Performance standards and evaluations in IR test collections: Vector-space and other retrieval models. *Information Processing and Management, 33*(1), 15–36.

Shortliffe, E. H., & Buchanan, B. G. (1990). A model of inexact reasoning in medicine. In Shafer, G., & Pearl, J. (Eds.), *Readings in Uncertain Reasoning*, pp. 259–273. Morgan Kaufmann, San Mateo, CA.

Shreider, Y. A. (1974). *What is Distance?* U. of Chicago Press, Chicago.

Singhal, A., Buckley, C., & Mitra, M. (1996). Pivoted document length normalizaton. In *Proceedings of the 19th Annual International ACM SIGIR Conference on Research and Development in Information Retrieval, Zurich, Switzerland*, pp. 21–29 New York. ACM Press.

Smeaton, A. F. (1984). Relevance feedback and a fuzzy set of search terms in an information retrieval system. *Information Technology: Research and Development, 3*(1), 15–23.

Sneath, P. H. A., & Sokal, R. R. (1973). *Numerical Taxonomy: the Principles and Practices of Numerical Classification*. W. H. Freeman, San Francisco.

Sombé, L. (1990). Reasoning under incomplete information in artificial intelligence. *International Journal of Intelligent Systems, 5*(4), 323–472.

Sparck Jones, K. (1972). A statistical interpretation of term specificity and its application in retrieval. *Journal of Documentation, 28*(1), 11–21.

Spink, A., & Losee, R. (1996). Feedback in information retrieval. In *Annual Review of Information Science and Technology*, pp. 33–77. Information Today, Inc., Medford, NJ.

Srinivasan, P. (1990). On generalizing the two-Poisson model. *Journal of the American Society for Information Science, 41*(1), 61–66.

Strzalowski, T. (1995). Natural language information retrieval. *Information Processing and Management, 31*(3), 397–417.

Svenonius, E. (1986). Unanswered questions in the design of controlled vocabularies. *Journal of the American Society for Information Science, 37*(5), 331–340.

Swanson, D. R. (1977). Information retrieval as a trial-and-error process. *Library Quarterly, 47*(2), 128–148.

Swanson, D. R. (1986). Subjective versus objective relevance in bibliographic retrieval systems. *Library Quarterly, 56*(4), 389–398.

Swets, J. A. (1969). Effectiveness of information retrieval methods. *American Documentation, 20*(1), 72–89.

Tanner, M. A. (1996). *Tools for Statistical Inference: Methods for the Exploration of Posterior Distributions and Likelihood Functions* (Third edition). Springer-Verlag, New York.

Tenopir, C. (1985). Full text database retrieval performance. *Online Review, 9*(2), 149–164.

Teugels, J. L. (1990). Some representations of the multivariate Bernoulli and binomial distributions. *J. of Multivariate Analysis, 32*(2), 256–268.

Turner, R. (1984). *Logics for Artificial Intelligence*. Ellis Horwood, Chichester.

Turner, R. (1990). *Truth and Modality for Knowledge Representation*. Pitman, London.

Turtle, H. (1994). Natural language vs. Boolean query evaluation: A comparison of retrieval performance. In *ACM Annual Conference on Research and Development in Information Retrieval*, pp. 212–220 New York. ACM Press.

Turtle, H., & Croft, W. B. (1991). Evaluation of an inference network-based retrieval model. *ACM Transactions on Information Systems, 9*(3), 187–222.

Tversky, A. (1977). Features of similarity. *Psychological Review, 84*(4), 327–352.

Tversky, A., Slovic, P., & Sattah, S. (1988). Contingent weighting in judgment and choice. *Psychological Review, 95*(3), 371–384.

Van Rijsbergen, C. J. (1986). A non-classical logic for information retrieval. *Computer Journal, 29*(6), 481–485.

Van Rijsbergen, C. (1974). Foundation of evaluation. *Journal of Documentation, 30*(4), 365–373.

Van Rijsbergen, C. (1977). A theoretical basis for use of co-occurrence data in information retrieval. *Journal of Documentation, 33*(2), 106–119.

Van Rijsbergen, C. (1979). *Information Retrieval* (Second edition). Butterworths, London.

Von Wright, G. H. (1957). *Logical Studies*. Routledge & Kegan Paul, London.

Watters, C. R. (1989). Logic framework for information retrieval. *Journal of the American Society for Information Science, 40*(5), 311–324.

Weiler, H. (1965). The use of incomplete beta functions for prior distributions in binomial sampling. *Technometrics, 7*, 335–347.

Williams, W. T., & Dale, M. B. (1965). Fundamental problems in numerical taxonomy. *Advances in Botanical Research, 2*, 35–68.

Witten, I. H., Moffat, A., & Bell, T. C. (1994). *Managing Gigabytes*. Van Nostrand Reinhold, New York.

Wood, F., Ford, N., Miller, D., Sobczyk, G., & Duffin, R. (1996). Information skills, searching behaviour and cognitive styles for student-centered learning: A computer-assisted learning approach. *Journal of Information Science, 22*(2), 79–92.

Yu, C. T., Buckley, C., Lam, K., & Salton, G. (1983). A generalized term dependence model in information retrieval. *Information Technology: Research and Development, 2*(4), 129–154.

Yu, C. T., & Salton, G. (1977). Effective information retrieval using term accuracy. *Communications of the ACM, 20*, 135–142.

Zwick, R., & Carlstein, E. (1987). Measures of similarity among fuzzy concepts: A comparative analysis. *International Journal of Approximate Reasoning, 1*, 221–242.

Index

$\mathcal{A}$, 112
  best-case
    approximate, 114
    exact, 114
  binary distributed terms, 112, 121
  binary relevance, 118
  continuous relevance, 118
  multivariate, 153
  normally distributed terms, 119
  Poisson distributed terms, 121
  relations, 193
  single terms, 112
  worst-case
    approximate, 114
    exact, 114
absorption, 59
agents, 4
analytic models, 2–3, 12–16
  verification, 14–15
  vs. experiments, 13–14
ASL, 78, 89–92
  best-case
    approximate, 154
    exact, 154
    single terms, 115
  bounds
    univariate and multivariate, 154
  multivariate, 153
  recursive computation, 164
  single terms, 114–117
  worst-case
    approximate, 154
    exact, 154
    single terms, 115
associativity, 59
asymmetric uniqueness measure, 85
autocorrelation, 217
autoepistemic logic, 200
average search length, *see* ASL

Bahadur-Lazarsfeld Expansion (BLE),
    135–136, 155, 166
Bayes' rule, 28
Bayesian networks, 138–139
belief function, 38
Bernoulli distribution, *see* binary distri-
    bution
best-case ranking, *see* optimal ranking
beta distribution, 35–36
  incomplete, 102
  prior for binary distribution, 35
biased sampling, 69
binary distribution, 24–25, 65
  multivariate, 130–141
Bochvar, *see* logic
Boltzmann chains, 146
Boolean logic, 57, 58, 172
Boolean retrieval, 57–60, 100, 171
  combined with vector retrieval, 62
Brookes, *see* E
browsing, 73–74, 169

Caitlyn, xi, 6, 178, 192, 213
cases, 218
certainty factors, 40–41
change points, 144
chi-square distribution
  inverse, 143
circumspection, 201
classification systems, 5, 73
closed world assumption, 199
collaborative filtering, 10
collocation, *see* term dependence
commutivity, 58
complementation, 59
conditional probability, 23
confidence interval, 117
conjugate prior distributions, 34–37
conjunctive normal form (CNF), 60, 189,
    190